THE ROBOT AND AUTOMATION ALMANAC 2022

FI THE FUTURIST INSTITUTE

Edited by Jason Schenker
Chairman of The Futurist Institute

THE ROBOT AND AUTOMATION ALMANAC - 2022

THE FUTURIST INSTITUTE

EDITED BY JASON SCHENKER

ISBN: 978-1-946197-76-4 *Paperback*
978-1-946197-77-1 *Ebook*

For futurists everywhere.

THE FUTURIST INSTITUTE

CONTENTS

CONTENTS

CONTENTS

PREFACE

FROM THE FUTURIST INSTITUTE

This is the fifth edition of *The Robot and Automation Almanac.* What began as a project of future history has become an important guide and touchstone of futurist thought to help shape and frame the expectations, opportunities, challenges, and risks presented by emerging automation technologies in the year ahead and beyond.

Technology has become more important than ever in the wake of the COVID pandemic. With supply chain and labor market challenges abounding, automation, robotics, and AI present tremendous opportunities. It also means that there is a greater need than ever before to think strategically about the future of these technologies. It is my hope that *The Robot and Automation Almanac - 2022* helps you consider the future of automation, robots, robotics, AI, and other related technologies.

On behalf of The Futurist Institute, I want to acknowledge and thank all of the contributors to *The Robot and Automation Almanac - 2022.*

As in the past four years, we have an amazing cohort of contributors in the fields of robots, automation, and AI this year. Thank you to Michael Walton, Bruce Welty, Shannon McGurk, Djamila Amimer, Jason Gouw, Bruce Bleikamp, Steven LaFevers, Louie Kangeter, Anton Larsson, Thomas Evans, Kaleb Steinhauer, Nawfal Patel, Shane Istre, Robert Handfield, Kevin Paramore, Martin Buehler, Lorenzo Carver, Jim Rice, Daniel Stanton, Patrick Davison, Crystal Parrott, and Mark Manning.

I am humbled by this list of amazing authors! Without them, this book would never have happened. Their visions of the future are both exciting and valuable. Their expectations for "the big thing" in the year ahead and beyond are important for understanding a wide range of technological developments, initiatives, opportunities, and challenges in the year ahead.

It is my personal pleasure to have been involved in this undertaking. The robotics, automation, and AI industries are still very new, but they are becoming more important. And it is with great pleasure that The Futurist Institute is able to support the companies in these industries with this almanac. I founded The Futurist Institute to help strategists, analysts, and executives become futurists in October 2016; we recently began our sixth year of operations.

I also want to thank The Futurist Institute team for providing support and feedback about this project. And I wish to thank Kerry Ellis, who designed a wonderful cover for the almanac.

With the inaugural edition of *The Robot and Automation Almanac* for 2018, The Futurist Institute created an invaluable tool for professionals, individuals, and investors seeking to understand the implications of robots and automation for their personal, professional, and investing lives. Multiple past editions provided unique insights and have become bestsellers.

We have similarly high expectations for how *The Robot and Automation Almanac - 2022* will presage changes in the year ahead, and that it will also help professionals think strategically about a variety of potential changes in the wake of the COVID-19 disruption that have loomed large over the economy, the supply chain, and the labor market.

The transition from the information age into the automation age is underway. Thank you for being a part of this dynamic shift.

The history of robots and automation is being written now — in this book. When people look back a decade from now, they will want to know how robots and automation evolved. They will want to understand how robots and AI became more integrated. They will wish they read *The Robot and Automation Almanac.*

Welcome to the future! ~

Jason Schenker FLTA®
Chairman of The Futurist Institute
Editor of *The Robot and Automation Almanac*

THE SCAVENGER HUNT ECONOMY AND THE CASE FOR AUTOMATION

Jason Schenker

- Chairman of The Futurist Institute -

Scavenger Hunt Economics

Business leaders, policymakers, and financial analysts always say that they want to see the economy firing on all cylinders. Indeed, it surely has been during 2021. The problem is that the economy has been running on eight cylinders — but it's only a six-cylinder economy.

Companies operating in the physical world have been subjected to what I call "scavenger hunt economics" over the past year and they could remain under stress to find the raw materials, goods, people, and means of transport necessary to operate successfully. Four fundamental factors have engendered these operational challenges that threaten to hinder profitability, despite a strong economy and high orders. These four factors include record high e-commerce demand, record high levels of goods demand, a tight labor market, and tight waterborne and on-road freight markets. These factors have engendered supply chain challenges and high inflation.

E-Commerce Demand[1]

For decades, e-commerce has been rising steadily. But during the COVID pandemic it soared in dollar terms — and as a percent of retail sales. Now, as vaccine distribution has reached scale in the United States, brick and mortar retail sales have increased to take back market share. But the absolute dollar level of e-commerce retail has continued to rise, and was at a new all-time high in Q2 2021 at around $222 billion, and fell rather modestly in Q3 2021 to almost $215 billion. In other words, e-commerce isn't going away. This is why any discussion of automation has to involve a discussion of the supply chain. Yes, there is a lot of excitement about the myriad activities robots are being developed to do. But the automation and robotics demands of the marketplace right now are unquestionably urgent in the supply chain. Moreover, robots do not take breaks, they do not go on vacation, and they do not get COVID. In order to continue to meet the demands of a supply chain never designed for single-piece batching, automation and robotics in warehouses, distribution centers, and across the supply chain are no longer optional. They are essential.

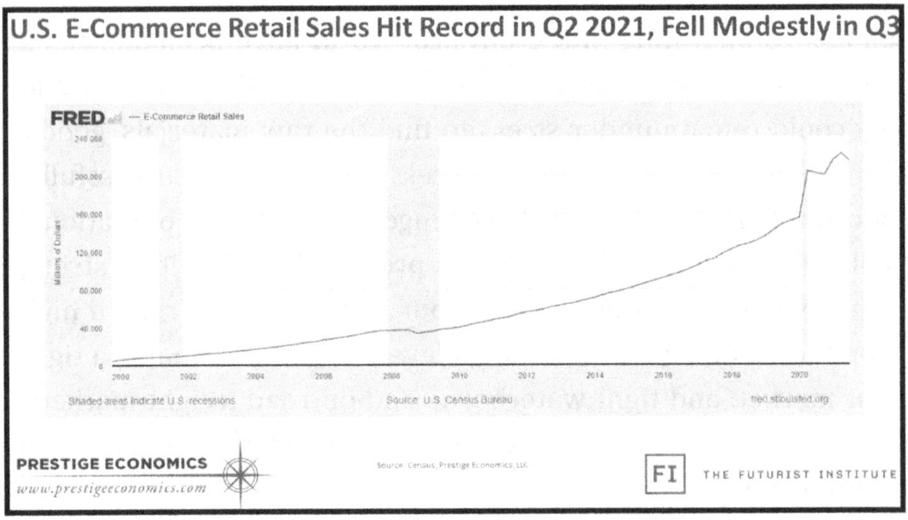

Labor Market[2]

The COVID recession was catastrophic for the U.S. and global labor force. But the response of both massive fiscal policy stimulus as well as monetary policy stimulus caused the sharpest bounce back in the U.S. labor force in history. The unemployment rate skyrocketed to 14.8% in April 2020, but fell back down to only 4.2% as of November 2021. Meanwhile, initial jobless claims — a measure reflecting the number of new people who lose their jobs in any given week — fell to a 52-year low in mid-November.

But there are signs the labor market may now be too hot. With record high real estate prices, record high equity markets, and an aging U.S. workforce, there is evidence that some would-be retirees have left the workforce permanently. Plus, the ability for people to work remotely has allowed some families to capture a massive geographic arbitrage that may also have permanently pulled people from the workforce. For manufacturing, supply chain, and other physical industries that were essential during the pandemic, the labor force tightness is even greater.

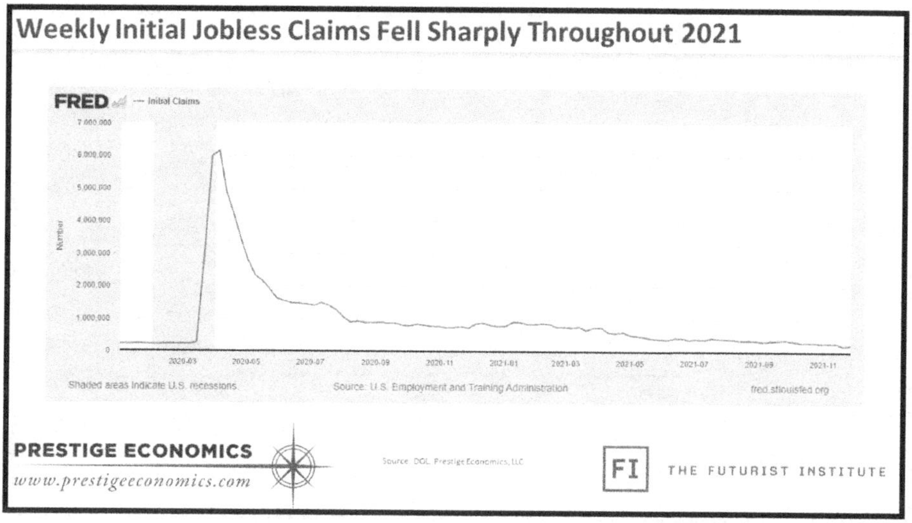

Producer and Consumer Inflation[3]

With such strong demand for e-commerce and other goods as well as a tight labor market, it should be no surprise that inflation has risen. For manufacturers, this inflation is particularly pronounced, with goods prices up +14.2% year on year through October 2021. And prices are likely to remain persistently high. The labor market tightness noted above is hitting manufacturers from every angle. There is a shortage of people to work in manufacturing. There is a shortage of truck drivers. There is a shortage of people to work in the supply chain and material handling industries. These challenges have compounding inflationary impacts at each stage of manufacturing, transport, and production. There has also been an impact of these factors on consumer inflation.

While producer inflation was up +8.6% year on year through October 2021, consumer inflation was up by a somewhat lower +6.2%. Although consumer inflation has been up less than producer inflation, consumer inflation in October 2021 was still over three times the rate the Fed was targeting before COVID.

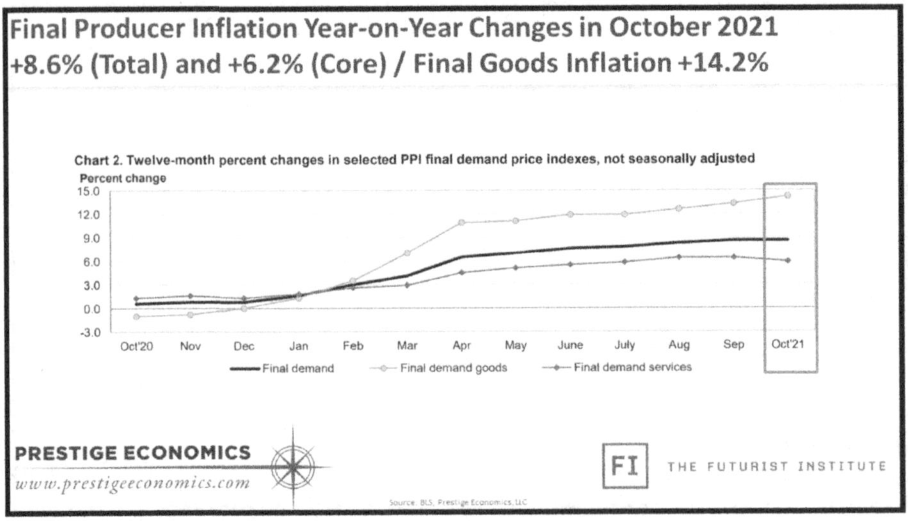

The Case for Automation

With all of the costs and stressors across the U.S. economy and supply chain, how can companies fill their record levels of new orders without risking profitability? After all, no one wants to have record orders and yet have trouble turning a profit because cost pressures rise more quickly than prices. There are only two ways to avoid this situation.

On the one hand, companies can try to push prices higher and faster for their goods and services. They are trying to do this now, but it's tough to raise prices quickly enough to keep pace. And no executive wants their company to lose market share because they raised prices too quickly. A brute-force approach to pricing also exacerbates the risks of what's known as a price-wage spiral. That's the rather unpleasant phenomenon, whereby prices rise due to strong demand and then wages rise, but then because of the higher wages demand for goods remains strong, and prices rise further. That, in turn, leads to more wage increases and so on and so forth.

But there is a more elegant solution. Companies can look to technology to help drive down costs. They can embrace the notion that a penny saved is a penny earned, and enlist automation, robotics, and data-enabled AI solutions to help reduce costs and drive up profitability, regardless of pricing in the marketplace. That can also remove some of the burden on executives to achieve profitability solely through forcing price increases at a pace quicker than competitors — or suppliers. But these projects often take time and require significant capital investments, and they are not without risks.

While the Fed Chair only retired the term "transitory" when referring to inflation in early December 2021, many in the manufacturing world had given up on the notion of inflation dissipating many months earlier.

The demands for e-commerce and other goods have just been too strong — and the supply chain issues currently plaguing U.S. companies, businesses, and consumers are not likely to dissipate with haste. Plus, labor market dynamics are poised to remain a challenge for industrial companies, especially as more workers retire and the tsunami of intellectual-capital-oriented jobs that allow remote workers to be anywhere offer compensation that undermines key pillars of profitability for some businesses that have relied on low-cost real estate and isolated labor markets for decades or more.

Despite the temptation to embrace a strategy of just chasing prices higher, there really is only one answer in the long run: technology. I wish that my assessment of these fundamental economic drivers were outliers, but they are not.

In the chapters ahead, you will read about the rising potential for robots and automation in the supply chain, the tightness in the labor market, the corporate focus on robotics ROI, the value of data to inform AI and boost profitability, and the growth of the metaverse outside the world of the physical. Real world physical challenges are the most pressing issues facing most businesses in 2022, and they will shape the demand and development of technologies in the year ahead.

The Case for Education and Education Provenance

In the longer run, one of the biggest challenges to the economy will continue to be human capital needs. This is why the topic of education comes up more than once in the chapters ahead.

The importance of education and the need to ensure the authenticity of human capital assets, like degrees, certificates, and certifications, is why The Futurist Institute partnered in November 2021 with iTRACE Technologies to create the joint venture NFT Degree, LLC.

Our mission is to create a way to verify the provenance of human capital assets by creating digital twins for physical education documents (like The Futurist Institute's FLTA® certificates). We do this by converting these physical documents into non-fungible tokens or NFTs that are recorded on a permanent blockchain. When paired with blockchain technology, the origin, non-transferable ownership, and record of these NFT education assets make their provenance permanent, immutable, and easy to authenticate.

Supply Chain and the Metaverse

Beyond human capital needs, physical supply chain needs — including the ability to confirm provenance — will also become more important in the years ahead. In some ways, it's why the metaverse is likely to thrive in 2022 and beyond. After all, the ability to satisfy the innate human desire to consume — our commodity fetishism — can also be more efficiently satiated in the metaverse.

Even when we live in a more immersive digital world, automation, robotics, and AI will continue to present outsize opportunities for both greater efficiency and sustainability. That only strengthens the case for automation and robotics in 2022 and beyond. ~

Jason Schenker FLTA® is the Chairman of The Futurist Institute, Co-Founder of NFT Degree, and the author of *Futureproof Supply Chain*.

ENDNOTES

1. U.S. Census Bureau, E-Commerce Retail Sales [ECOMSA], retrieved from FRED, Federal Reserve Bank of St. Louis; https://fred.stlouisfed.org/series/ECOMSA, December 6, 2021.
2. U.S. Bureau of Labor Statistics, Unemployment Rate [UNRATE], retrieved from FRED, Federal Reserve Bank of St. Louis; https://fred.stlouisfed.org/series/UNRATE, December 6, 2021. And U.S. Employment and Training Administration, Initial Claims [ICSA], retrieved from FRED, Federal Reserve Bank of St. Louis; https://fred.stlouisfed.org/series/ICSA, December 6, 2021.
3. U.S. Bureau of Labor Statistics, Producer Price Indexes, retrieved from BLS; https://www.bls.gov/ppi/, December 6, 2021. And U.S. Bureau of Labor Statistics, Consumer Price Indexes, retrieved from BLS; https://www.bls.gov/cpi/, December 6, 2021.

OPPORTUNITIES AND LIMITS OF ROBOTS AND AUTOMATION

Bruce Welty

- Technology Executive and Entrepreneur -

The Case for Robots with ROI

Robotics and automation industries are likely to see upheaval and change including the future consolidation of successful companies, and the weeding out of failed ones. We are likely to see this play out more seriously in 2022 as successful companies come together and win market share. In the end, companies that are unsuccessful will disappear.

This process of creative destruction is unlikely to completely play out in 2022. It will likely take many years. But the process is likely to accelerate in the year ahead. The most vulnerable companies will be those that are not ROI focused and are too busy chasing headlines and flashy images to focus on the tangible bottom line.

People spend a lot of time focusing on the theoretical value of robots doing things as humans do rather than considering tangible business cases. Consider the humanoid robot waving a checkered flag at a car race: it boggles my mind that that is where we want to spend our time and money and limited roboticist resources. When it comes to bipedal robotic potential, I'm a bit of a robot pessimist.

The truth is that robots are really good at what people are not good at, and that is where roboticists should be focusing their efforts.

Taking Warren Buffet as a muse, we will figure it out. In a capitalism system, we find ways to work it out. We have some real challenges in the world today, and the focus for fixing problems in the here and now is on robots. Automation and robotics can help resolve those issues. The great filtering of companies in this industry will favor companies that solve problems and add value. The market will fix it.

In 2022 and the years ahead, it will become increasingly important to focus on exactly those areas where practical advances can make significant contributions to productivity, business operations, and the economy.

The Labor Force Imperative

The imperative for practical ROI-centric robotics and automation systems will be greater in 2022 than in past years. There are additional pressures that companies in the space will experience in the year ahead, including significant headwinds from rising labor costs, tight labor availability, rising customer expectations, and growing competition.

There are labor market issues in the supply chain. Part of the issue is that these are hard jobs and people don't want to work in supply chain jobs like trucking and ports. Some warehouses had trouble filling up well ahead of Christmas 2021, with some shipments ending in early November due to a breakage in global supply chains. As a result of these dynamics, businesses have begun considering a switch from just-in-time inventory management to just-in-case. Operational focus for many companies is shifting away from keeping inventories at minimums to maximize profit. Now, the focus is to more fully consider the risks if they run out of stuff. However, it is much more difficult to manage a full warehouse.

The supply chain is just that: a chain. When one link in the chain breaks, there is disruption that cascades through industries and across the economy. Those breaks could incentivize greater demands for automation and robotics. After all, we need to maximize cubic volume of a facility, and we need to use space better. This is why we need — and are likely to see — more automation coming from people who understand robots and fulfillment. It's inevitable given current labor market conditions, and secular demographic changes that lie ahead.

While there are supply chain issues, big pushes into e-commerce have also exacerbated the process of getting material to the shelves. There is still more to shake out in retail, like who can get the products and parts and how do they make them available to demand.

In the traditional brick-and-mortar retail economy, consumers provide massive amounts of free labor as they pick, pack, and ship all of the goods they purchase at stores. With the growing popularity of e-commerce people don't want to drive around and do these things for free anymore. In short, the labor burden is shifting from consumers to retailers and logistics providers on a massive scale, and the labor market cannot support this rising demand.

Looking ahead to 2022 and beyond, as supply chain businesses across the economy try to reduce pressure on ports and warehouses, I think we are going to see tremendous growth in warehouse automation to close the growing labor gap.

Of course, there are limitations to automation as well. Too often, there is a trade-off where more automation leads to less flexibility. This is why companies like Locus provide automation equipment that can add flexibility. And it is in that balance between automation and flexibility where an important middle ground of opportunity exists for automation in the supply chain. The automation is needed, but so is that flexibility. By offering both, an automation system can be both important and valuable.

Robotics Development in Context

For automation and robotics, right now feels to me like its 1985 in the computer business. Back then, the world of personal computers was changing rapidly. Some of the biggest names then no longer exist today. Similarly, some of laggards then have become some of the biggest PC companies today. The landscape of companies is changing quickly, and there are filtering and seismic shifts ahead.

There are some critical fundamentals that will shape the various permutations of robotics and automation that dominate our lives. Fundamentally, material science is hard, physics is a real thing, and edge case ennui can discourage engineers working on critical products and systems.

For widespread adoption of self-driving vehicles, in particular, there are some major issues to be resolved. Some of this is around technology, but other parts are related to regulation, insurance, and liability. For self-driving vehicles, a key question to answer will be, "Who is responsible when an accident happens?"

The Need for Fault Tolerant Machines

People won't easily accept that code might cause an accident and kill someone. One practical coding element of fault tolerance that is particularly challenging — but necessary — for self-driving vehicles to succeed is an operational reset. Every technological product should have a way to reset its fault tolerance. For robotics and automation, there needs to be a ctrl-alt-delete function.

For personal computers, ctrl-alt-delete was developed as a way to reset operations. It gave users confidence that no matter what went wrong, there would be an escape hatch. It made personal computers operable by any one, no matter what. In the context of reliability, it made computers reliable. This is an important concept for all computerized and automated systems, including those we developed at Locus.

As we consider the future of automation, there is a great deal of focus on self-driving vehicles. The same kind of fault tolerance reset provided by ctrl-alt-delete is a necessity. You need a ctrl-alt-delete function so that no matter what happens, anyone can reset it without any skills. But how do you create that kind of fault tolerance for a car going 70 mph? That is a critical engineering challenge.

You can turn off autopilot as a reset and have the vehicle tell its driver to take control, but that's not autonomous driving. That's human-assisted driving, and while it might avoid an accident, it doesn't remove the need for a human driver and it is unlikely to save any money.

Every product should have a ctrl-alt-delete, but I don't see that with autonomous vehicles yet. Maybe we will get there someday. We probably have to. After all, what's the alternative to ctrl-alt-delete? Thirty-five years on, we don't have one. ~

Bruce Welty is an entrepreneur and business builder in the robotics space and the founder of Quiet Logistics and Locus Robotics with valuations in excess of $2 billion.

AUTOMATION ROI IS POISED TO ACCELERATE

Michael Walton

- Microsoft Industry Solutions Executive -

COVID-19 has been an accelerant for automation technologies, and there are three major trends in the wake of COVID that are likely to support a further acceleration of automation in the year ahead. These include labor market tightness and the great resignation, democratized and standardized technology, and the push for sustainability. These key trends are occurring not just in the United States, but in regions all over the world.

The Great Resignation.

One of the biggest trends impacting the economy right now is the so-called great resignation. People are changing jobs and the labor market has become extremely tight very quickly. This level of transition, as well as the associated higher labor costs, are likely to further incentivize employers to deploy automation in the year ahead.

Right now, you have high consumer demand for many products such as fast food and consumer electronics. For those products to be made and sold someone must cook the food, conduct maintenance in plants, do janitorial duties, and every other kind of job. At the same time, workers are demanding far more money to do those jobs, resulting in a dramatic increase in cost. Existing automation technology is about to be rolled out at scale. For years, we have had the ability to automate many kinds of roles, but the ROI (Return on Investment) wasn't there. As labor costs now rise dramatically and many jobs go unfilled for long periods of time, the ROI of automation technology is becoming more viable.

This is something that has happened in the economy time and again. People used to hand wash their clothes. Then, automation came out and most people didn't want to spend their time washing clothes anymore, nor did they want to be paid to do that service by hand using a manual method. Naturally, people accepted automation as a must have. Of course the "washing clothes" use case seems obvious, but I am not sure it was so obvious at the time it happened.

Another example is robots that sweep floors (now robot vacuums are becoming common place). It's a big help and people don't want to go back to cleaning their floors the way they used to. While the example of the robot vacuum is a consumer good, there are even greater use cases for widespread automation and robots in the wake of COVID. After all, we aren't just talking about the pure labor ROI of automated tasks.

The pandemic, health concerns, and a preference for contactless technologies are likely to engender a further acceleration in the adoption of automation across the gamut of roles. Paired with inflation (resulting from increased labor costs and supply chain costs) and job vacancies, these dynamics make automation not just reasonable but necessary! You can't pay people at fast food restaurants a lot more than before COVID and expect that there won't be business, corporate, inflationary, or economic consequences that everyone else must endure.

Whether we are talking about a taco or a pizza, it is probably cheaper, safer, and more efficient to make those food items with the right automation than with humans. Companies and consumers have a breaking point for costs. Many of these jobs will mostly — not entirely — get automated out sooner than later, because the human capital costs will be too high.

This is also true in manufacturing, which saw exceptionally strong demand over the past two years as consumers shifted their consumption to goods away from services. Warehousing, transportation, and logistics are likely to see much more automation — especially in controlled environments for high manual labor jobs.

Because of high human capital costs, we are likely to see hard and software bots (leveraging Artificial Intelligence) more frequently used for things like cycle counting inventories, cleaning, picking, pallet moving, and even delivery. For example, take the cycle counting use case.

Many manufacturers either outsource cycle counting or pay some kind of labor to do this. During COVID this become even more difficult for obvious reasons. Cycle counting is a very manual, labor intensive, and time-consuming process. Now, you can use a robot or drone that takes uses a camera and takes photos. With AI you can process those photos to tell the system how much inventory there is. It can be on wheels or in the air. No matter how you are deploying that drone, the accuracy is there. It has become truly fantastic. Humans can work exceptions and investigate as needed, but humans are not the first line of defense — or at least they won't be much longer. I have worked with many customers who do this very thing today and it wasn't that difficult.

Democratized and Standardized Technology

More technology is not sufficient to ensure success in automation. The tech must be accessible, which is why there has been a democratization of automation, remote, and data technologies. Companies have standardized the software and technology at scale to make this happen, so that robots and automation are easier to deploy, operate, and maintain. Thus, the cost of automation overall is improving while the cost of human labor is drastically increasing!

This new metaverse including automation working at scale depends on standardization for it to be scalable and maintainable. And the communication piece is being standardized globally. Microsoft, Google, IBM, and AWS are all working on this issue. I believe that Microsoft is ahead. But that's largely because we were the first mover.

Because of Microsoft's passionate approach to open source and global standards adherence, I believe other companies are going to follow these same standards for communication. In short, the technology has been democratized and standardized — and that allows for scale.

This solves real problems for customers. For example, a discrete manufacturing company that has worked with us using robots that incorporate AI with cameras, as well as this democratized communication standard to collaborate with humans, is doing the work that once was thought could not be done. The results speak for themselves. Takt times that could previously never have been reached are now being achieved without compromising quality! The cameras backed by AI act like eyes of robots and humans. They are able to ensure humans adhere to work instructions far faster, because the robot is able to adhere in real time to its mission. Now, the robot can do that work even faster when working with a human. This is important for responsible AI, where humans play an important role in this dialogue. By the way, this is not just coming, it's here. So, expect to see and hear a lot more about it in 2022.

An engineer sitting anywhere in the world can design how to make something and those options can change in real time. These goods with semi-custom options need to be assembled with different parts and specific instructions. In order to have the same line create different goods, you have to take the instructions from that engineer to the robot and have the warehouse bring the parts.

It's difficult to program a bot to do everything at all times. But with AI and custom vision, now we can get there. Humans collaborating with the bot can put the parts in the right place or confirm the right parts. This accelerates the process, with humans doing more supervisory work.

Things aren't made the Henry Ford way anymore: one product on one line. Now, many items can be made on one line where humans and robots are working together.

Historically, there were higher than acceptable scrap rates and rework rates, often caused by nonadherence to the work instructions, resulting in errors like incorrect assembly or use of the wrong parts. Now, because bots are working with humans, there can be significant improvements to the process in a way that boosts efficiency and also lowers scrap and rework rates. This means that you make more per hour, reduce your costs, and increase your quality. But the only way you can do that is by leveraging AI and automation, which enables more complicated robots to work with humans.

Automation that integrates AI working and collaborating alongside human capital is both here and real! It is not fiction anymore. Most importantly, it's the integration of AI with automation that allows for more advanced robotics capabilities. In short, with AI and human collaboration, robots can do more. This will be much more commonplace in 2022 and beyond.

Sustainability

Sustainability is the third critical element that will support the next level of automation in the year ahead. After all, automation makes things more efficient. You can track what bots are doing to control more and see more. This transparency is likely to result in dynamics where companies have a greater chance of achieving carbon neutral operations. And the more waste and scrap you remove, the less fuel you use. The use of automation is likely to result in more optimized and efficient operations that are a benefit for humans as well as a lever of sustainability.

Acceleration in the Year Ahead

Automation decisions are business decisions. With a rising value proposition and ROI, automation adoption is likely to accelerate in the year ahead, potentially reaching a fever pitch as companies struggle to augment their labor forces, contain inflation, and boost their productivity. The standardization of technology that supports automation will make it easier than ever to deploy and operate automated solutions. Finally, institutional and activist investors are likely to underwrite the transition to automation with financial mandates to capture sustainability benefits. The case for automation in the year ahead is truly compelling. ~

Michael Walton is Director and Industry Solutions Executive at Microsoft, where he is helping discrete manufacturing companies to digitally transform.

WANTED:
RELIABLE ROBOTS (RIGHT NOW)

Thomas Evans

- Robotics Chief Technology Officer at Honeywell -

Demand Is Rising for Fast, Efficient Automation Deployment

While interest in automation has been increasing in distribution and fulfillment (D&F) operations for several years, it's stronger than ever as we head into 2022.

Several key factors are driving the growth. The first is unprecedented demand for e-commerce, which is far more than a short-term blip caused by the recent pandemic. A related challenge is the industry's growing struggle to find enough labor, made worse by high turnover rates. A third trend that will help offset both of these problems in 2022 is a combination of several technological advances — major leaps forward that have enabled robotics to go beyond their traditional role as pre-programmed automatons practical only for predictable, repetitive tasks.

Today's smart robots are achieving an unprecedented level of independence, able to navigate ever-changing environments and solve many workflow challenges on their own.

The impact of these changes will be significant and wide-ranging. Over the next two years, industry experts predict that half of all organizations in the supply management sector will invest in applications that support artificial intelligence (AI) and advanced analytics capabilities. A growing number of distribution center (DC) operators no longer talk in terms of "if" they will invest in automation, but "when."

In fact, many *aren't* waiting. According to the Association for Advancing Automation, sales of robotics were up 37% in 2021 compared to the same benchmark in 2020, reaching nearly $1.5 billion. Earlier this year, Google Cloud research found that two-thirds of manufacturers who are using AI are relying more heavily on it. What's more, it's becoming clear that those who are making the transition sooner are enjoying a competitive edge that could quickly leave modern Luddites behind.

This new paradigm suggests that the top trend in robotics for the coming year won't be any single technology, because *multiple* innovations are simultaneously transforming the way we think about robotics in DCs. Instead, we foresee a drive to find partners who can demonstrate a capability to increase the efficiency, uptime, and reliability of robotics — and integrate them into a total solution — while deploying them faster than anyone else.

E-commerce Demand Is Hotter Than Ever

The explosion of e-commerce was already well under way prior to worldwide lockdowns. While the pandemic didn't create these trends, it dramatically accelerated changes that were already in progress. In addition, many consumers who were formerly resistant to shopping online not only learned to trust and become comfortable with the process, but discovered the convenience of doing so.

These and other factors mean e-commerce growth won't go away as the pandemic subsides — or any time in the foreseeable future. In 2022 alone, online sales are projected to hit an all-time high of $1 trillion.

Manual labor simply isn't sufficient to keep up with these levels of rising demand. And that's *if* you can find enough workers — which remains the number one challenge for nearly every DC operator in the business. The most current statistics we've seen indicate that the warehouse industry now employs over 50% more workers than it did 5 years ago. Even though 13.6% of that growth was in the last year alone, it's still not enough. The warehouse and transportation industry had a record 490,000 openings in July, a gap experts predict will continue to widen.

Another growing concern is that the DC and warehouse workforce may be aging. *Some companies report that most applicants are in their 50s and 60s,* and those younger than 30 rarely stay more than a month. Worst of all, the warehouse industry has already cycled through millions of workers, many of whom say they've sworn off warehouse jobs.

Altogether, the challenges of today's e-commerce businesses would be more than enough to explain why interests in automation and robotics are reaching all-time highs. But it's only part of the story.

The Technology is Here Today

Many D&F operations have been on the fence about robotics for years. Although they've long since recognized the need for automation, they've expressed concerns that the technology wasn't quite ready for the changes DCs deal with every day.

From keeping track of the constantly changing physical locations of products, to interpreting virtually limitless pallet stacking patterns and new packaging types, to adapting to peak selling seasons and seismic shifts in consumer demand, no DC can afford to shut down and reprogram its robots every time the market changes.

Simultaneous advances in multiple technologies are enabling robotics to overcome these fears. Today's smart robots offer a game-changing level of sophistication that enables them to solve problems on their own, adapt to changing conditions without costly reprogramming delays, and provide long-term, "future-proof" performance.

We don't have to wait for the breakthroughs that will make this possible to be developed sometime in the future. We have them *right now.*

Key recent advances include:

Advanced Logic — Machine learning (ML) and AI are making robots smarter and faster, enabling them to thrive in dynamic DC environments. Modern control logic makes robots capable of accommodating the unique challenges of the task at hand, solving problems, and optimizing movements to ensure maximum speed and efficiency.

Computer Vision — Advanced vision and perception technologies are playing a critical role in empowering robots to operate autonomously and interact with their surroundings. These technologies provide the sensing functionality that enables robots to identify objects, plan movements, and respond to challenges like jams, unrecognized items or obstacles. They're also vital to the safety of human co-workers, vehicles, and more.

Mobility — Autonomous mobile robots (AMRs) continue to expand their footprint, freeing up scarce human workers by taking on repetitive transport jobs. AMRs are an increasingly popular starting point for automation because they can be deployed quickly — often in a matter of hours — and easily scale up or down to accommodate peak volume or changing needs. Many transport jobs traditionally performed by a worker with a pushcart, pallet jack, or forklift can now be handled by an AMR with little or no human supervision.

Simulation and Emulation — These synthetic data modeling strategies for both greenfield and brownfield operations are enabling even further optimization. Dozens or even hundreds of possible scenarios can be tested in a matter of hours, identifying the most efficient designs and processes for a specific site's needs before any physical equipment is ordered.

Streamlining Deployment

As exciting as these technologies have become, DCs can't just change their operations overnight by installing a few robots. What's more, it's no longer enough for developers to "throw a product over the fence" with a manual and hope for the best. Getting the most out of today's automation technology requires deployment strategies at least as intelligent and efficient as the robots themselves.

One of the most effective ways to ensure smooth and successful deployment is to build overlapping layers into the various teams. Creating links among the research and development (R&D), project engineering, field installation and support teams is a proven way to eliminate silos of expertise, ensure clear communication, and minimize commissioning delays.

Remote interfaces are also becoming more popular as automated solutions take on bigger challenges. When a robot or system has an issue, it can use the remote interface to call for human backup. This capability provides the dual benefit of clearing the current problem while updating the ML model. The robot will be smarter the next time it encounters a similar problem, while scarce workers can provide value in more satisfying roles.

These are just two of the time- and cost-saving strategies an experienced integration partner can provide. While many vendors sell automation solutions, it's best to work with an organization recognized by the Association for Advancing Automation (A3) as a Certified Robot Integrator.

To thrive in 2022, you'll also want to look for expertise that goes beyond the robots themselves. Seamless integration of robotics into a complete ecosystem can enable D&F operations to see every aspect of the big picture in one place, from order receipt and release to picking, packing, and shipping. Equipped with the abilities to track, trace, and optimize product flow throughout the DC, you can significantly improve inventory and order accuracies to raise your operation's service levels.

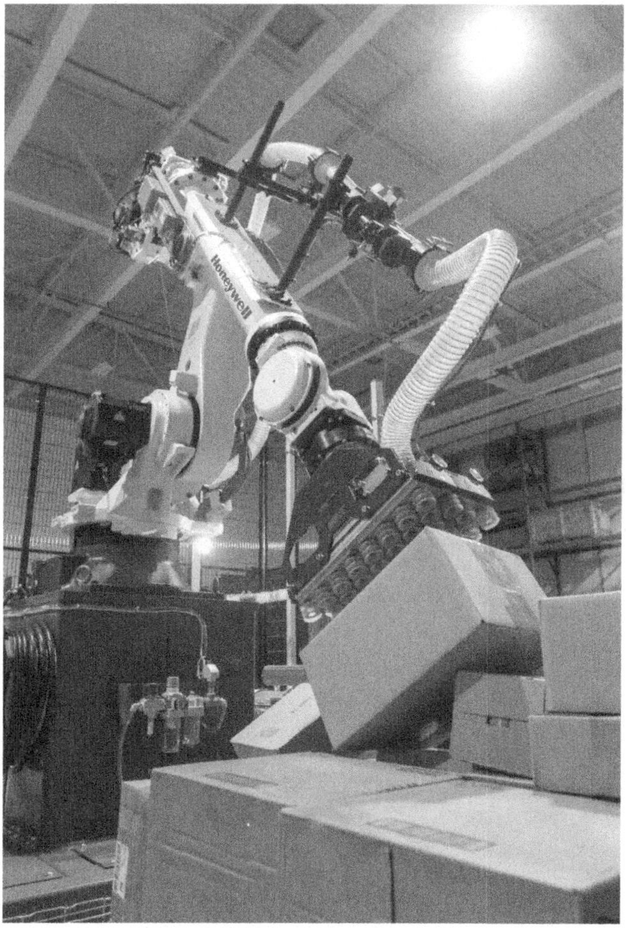

To achieve these goals, your integration partner should have an end-to-end knowledge of the entire D&F ecosystem, be able to demonstrate the measurable results you can achieve with automation — including the system's future potential — and provide full support capabilities. ~

Dr. Thomas Evans is the robotics chief technology officer for Honeywell. He is responsible for providing the technical vision and development for robotics and AS/RS shuttle solutions. Dr. Evans has twenty years of experience in robotics and engineering research. He holds a Ph.D. from West Virginia University.

WHY ROI IS NOT THE FIRST CONSIDERATION IN AUTOMATION

Patrick Davison

- Managing Executive of The Robotics Group at MHI -

Crystal Parrott

- VP of Engineering at Plus One Robotics -

If you have been considering automation solutions for your material handling operations over the past few years, you have likely spent quite some time reviewing online articles, webinars, and resources discussing the opportunities and challenges associated with implementing automation solutions. You have probably also noticed that several of these resources almost immediately discuss return on investment (ROI), and how achieving a satisfactory ROI is essential to launching any automation project.

The preponderance of ROI discussions could lead you to believe that an automation project starts with a proposal that achieves an acceptable ROI, then proceeding with the project.

The Robotics Group (TRG), an industry group of MHI, was formed in 2019 to inspire confidence and foster the adoption of robotic solutions to drive efficiency in supply chain applications. TRG has produced a "Buyer's Journey" framework that provides a six-phase roadmap for practitioners to follow when considering a greenfield or brownfield automation project after the end user has clearly outlined internally why they need to consider automation. Of course, the TRG Buyer's Journey includes ROI considerations, but expands upon five other important phases:

Business Case: Why do you need to consider changing your current operation?

Goals: What are the operational goals your company is hoping to achieve?

Process Improvement: How can operations be modified to achieve your goals, and how can automation assist various operational tasks?

Evaluation and Selection: Can the solution achieve your goals and deliver an acceptable ROI?

Implement: What will lead to successful goal attainment?

Sustain: What will the new process require after installation in order to fulfill the operational goals over its life cycle? Can the system be expanded or modified to accommodate future needs?

These phases are discussed further below.

<u>Business Case and Goals:</u> At the onset of a project, the team should consider everything it hopes to achieve by implementing automation at the facility. All too often, the primary reason given for material handling automation is to serve as a substitute for labor, which is deemed either too unreliable or too expensive given current business conditions. Apart from labor savings, goals could also consider the following operational factors:

Operational – Improvement in capacity or throughput: How much capacity would the system need to process? How many exceptions would the system need to be addressed, and how will they be handled?

Facility – Reduction in size or added flexibility: How much space will the system require?

Integration – Ease of adaptability with external components: How will the new system integrate with the current enterprise? Will it be adaptive to future revision or expansion?

Energy – Reduction in usage or environmentally beneficial: How much energy will the new system require? Does the facility have adequate electrical or pneumatic capacity to implement the new system?

Financial – Meets or improves metrics: What is the project budget? What are the ROI goals for the system?

Process Improvement: When automation is implemented into an existing operation, it could require that the operational processes be modified to accommodate the automation. There could also be a desire to increase throughput or efficiency, or to achieve other goals. The project team should consider ways in which the existing processes can be improved to achieve the goals.

Consider Alternatives: At this point, the project team should consider tasks that could be automated or where automation could assist or enhance the process. For material handling operations, there are several different types of automation products that can be implemented, including automated sortation & retrieval systems (ASRS), automatic guided vehicles (AGVs), autonomous mobile robots (AMRs), "smart" conveyors, and industrial and collaborative robots. These systems can be implemented to function either without human intervention or to work alongside humans to enhance the work experience. Pilot projects could be considered to evaluate proof-of-concept or to test how the process modification and automation would function.

Evaluation and Selection: At this phase, the project team ought to be able to put forth a proposal, outlining operational goals, process improvements, and recommendations for capital expenditures. The project team should be able to assess whether the proposed system would meet the business case and goals. But if the goals aren't met, the project team should be able to prioritize how it intends to address any deficiencies.

The enterprise can then decide whether to proceed with the project. In essence, ROI is a gate that opens when the project team has successfully addressed the business case and goals with automation solutions.

Implementation: After the selection process is complete, contracts are awarded, and the system can be installed and implemented. It is worthwhile to evaluate the various steps of installation, implementation, and deployment to ensure the accuracy of the evaluation and selection process — and to ensure that your goals are achieved.

Sustain: Most systems require maintenance or modifications over time to ensure proper function. Training could be required for employees to understand how to utilize or maintain the new system. The enterprise should prepare to address ongoing expenses associated with the system throughout its life cycle.

What Can Go Wrong?

There are instances where automation deployment is not successful. A few such instances are described here using the Buyer's Journey framework, but are all due to a breakdown in the process or unachievable goals:

Select then Evaluate: This occurs when a company selects a vendor or automation solution before clarifying its business case or defining its goals. Perhaps the company has "use it or lose it" budget or are swayed by a colleague or an outside sales representative.

The automation device is delivered to the facility without adequate context of what it is supposed to achieve, how it will fit within the existing processes, or evaluation of alternatives. If this occurs, it is likely worthwhile to reflect back upon the Buyer's Journey to ensure that the first three phases are considered before implementation.

"Pilot Purgatory": This occurs when a company is forced to iterate multiple times through the first part of the process without success. This is called "Pilot Purgatory" because the project gets stuck in a "do-loop" and never proceeds through completion. Ultimately, Pilot Purgatory occurs when the pilot does not meet one or more goals, requiring the project team to either evaluate other options — or revise its goals or processes.

Vague ROI: There is no shortage of opinions on what should be included in an ROI calculation. All too often, the ROI calculation for an automation solution involves a substitution of automation for labor, with the calculation assessing whether the capital expenditure can justify the savings from reduction in labor.

There are ongoing discussions on all of the components of labor, including turnover, absenteeism, training, human error, etc., all of which make automation more attractive. Conversely, the relative inflexibility of automation and handling exception cases make automation less attractive.

"Close Second": When it comes to implementing automation in material handling applications, many practitioners take a risk averse approach. Instead of taking a leading position and implementing novel automation solutions, they prefer instead for other companies to establish and implement solutions and will continue on their journey only when technologies are deemed to be well established. In essence, they let their competition complete their Buyer's Journey, then start theirs after someone else has proven the technology to be successful.

When properly implemented, automation solutions can help companies achieve their operational goals and fulfill their mission. Taking an iterative, process-oriented approach can help companies mitigate operational risks and select and implement automation solutions in a manner that helps achieve its operational goals. Short-cutting or avoiding phases can lead to unsuccessful launches and cause companies to shy away from implementing similar projects in the future.

Plus One Robotics system handling bulk parcel with a Yaskawa robot.

The bottom line is use of the TRG Buyer's Journey framework helps companies succeed in deploying automation solutions, including those with novel solutions. The key is keeping the organizational goals in mind when identifying where automation and technology fit into the overall process. ~

Patrick Davison is the Managing Executive of The Robotics Group and Director of Standards at MHI, where he is responsible for administering the development of over 30 American National Standards for material handling technologies. He was formerly the Director of Standards Development at the Robotic Industries Association.

Crystal Parrott is the Chair of The Robotics Group and Vice President – Engineering at Plus One Robotics, where she leads the engineering organization and spearheads the product development of automation solutions for e-commerce fulfillment, warehouses, and distribution centers.

THE IMPORTANCE OF SUPPLY CHAIN FLEXIBILITY

Daniel Stanton

- Author of *Supply Chain Management For Dummies* -

Jason Schenker

- Chairman of The Futurist Institute -

Making Technology Work for the Supply Chain

In 2020, the approach for many companies was to automate everything possible, and to deal with integration later. In 2021, companies made significant investments in automation, robotics, and AI with a greater focus on ROI. Now, looking ahead to 2022, the most important dynamic we expect is for companies trying to integrate their investments to maximize and elevate ROI, reduce friction, and increase flexibility.

This multi-year evolution represents a significant change from tactical reaction to deliberate strategy. We see tremendous opportunities for companies that can de-silo their supply chains and harness the capital assets and technology at their disposal.

From the Leap Forward to a Leap Together

Part of the evolution to integrating operations and shoring up ROI in the year ahead will require a shift from the COVID sprint forward to a collaborative leap together in which friction is reduced and people are likely removed from parts of the system.

Because integration of physical and digital technologies across operations involves complicated planning, execution, and visibility, companies looking to benefit from these changes will need to perform gap analysis to identify the leverage points where new investments in technology can provide the most value.

Often, human-machine interfaces present inefficiencies, so the big gains may come from identifying processes where humans limit efficiency and constrain supply chain operations. This could be in the form of robotics, where machines perform physical activities. But it may also be in the form of algorithms that replace human judgement and decision-making for business processes.

In years past, there was significant concern about automation leading to unemployment in both blue collar and white collar jobs. Fortunately, goods consumption has been so strong in the wake of COVID that automation and robotics do not currently present a joblessness risk in supply chain industries. In fact, the labor force is so tight that there is a persistent lack of unemployed, available workers in the industrial and retail sectors. As such, more automation and robotics in the supply chain could boost output and productivity.

Of Bottlenecks and Bull Whips

In the year ahead, the cartography of supply chains will be changing. There are always risks across the supply chain, but companies should now be paying more attention. This means that strategists, analysts, and executives planning the security of their supply chains in the year ahead will need both a roadmap and a futurist framework to address evolving bottlenecks, shortages, and even geopolitical risks.

Just like supply chains and geopolitics, technology is always in flux. The mounting complexity and risks, against a backdrop of revealed bottlenecks, should force an honest evaluation of sourcing parts, equipment, and labor. But will it? Or will these risks, and the attempt to address them, merely be paid lip service, rather than genuine strategic analysis and planning.

The Case for Flexible Automation

Because of the bottlenecks and other pervasive supply chain challenges, many companies will no doubt seek to embrace automation in the year ahead. One prudent strategy that many companies should pursue is that of flexible automation. There is a need to use technology as a lever of flexibility within supply chains, because shortages that lead to surpluses have been part of a cycle that has repeated itself time and time again across business cycles.

An important lesson from past cycles is that there is a significant risk that we could see the supply chain overcorrect in the near future as well.

The question of timing is difficult to nail down. Different goods and different commodities will probably experience surplus supplies just as demand is falling. But it is difficult to predict exactly when this will occur for any type of good or commodity. The timing depends on complex factors ranging from inventory levels, to consumer behavior, to product lifecycles. Rather than trying to accurately forecast the timing, a better approach may be to build flexibility into the supply chain that can be toggled up and down to adapt to emerging economic realities. This would allow companies to produce and ship more (or less), depending on the supply and demand dynamics.

For industrial manufacturers, supply may not be able to keep up with demand. There are also risks of shadow orders exacerbating supply chain issues. Companies desperate for commodities or goods may engage in gaming by placing multiple orders for the same items, knowing that they can cancel some of those orders in the future. This behavior, which is common during times of shortages, results in "phantom demand" and surplus supplies, which in turn amplifies the bullwhip effect.

When the bullwhip cracks, the supply chain is left with a hangover. The cure is to replace inventory with information. Better data and collaboration, combined with AI and more flexible automation in the supply chain, could enable predictive analytics and help to mitigate the glut of goods. An important note of caution: as supply chain technologies become more integrated, companies need to manage data quality and integrity or else their data could create garbage-in / garbage-out dynamics and contribute to the big swings in the supply chain bullwhip.

Just-in-Case Inventories Versus Just-in-Time

Inventory management approaches may also change in 2022. Just -in-time is an inventory management strategy that was pioneered by Toyota and has been widely embraced as the key to keeping companies lean and their operating costs low. But lean inventory approaches rely on having a responsive and reliable supply chain. When you do not, or cannot, trust the supply chain, just-in-time becomes risky.

In the wake of COVID, the stresses on U.S. and global supply chains left many orders unfilled, because not enough goods or materials were on hand. The manufacturing and distribution nodes in the supply chain simply could not respond quickly enough to the changes in demand. As a result, companies and individuals have lost trust in the supply chain system, and they may strategically decide to keep more inventory on hand in the future — just in case there are shortages or a surge in demand.

The supply chain issues of 2021 are likely to cast a shadow on the years ahead. Consumers impacted by the shortages and stockouts are likely to keep more goods on hand than before the pandemic. For companies, these issues are likely to foster an expectation that they hold just-in-case inventory, along with their just-in-time inventory.

Just-in-case inventory may make companies and individuals feel safer and more confident, but it is not a capital-efficient strategy. The months' worth of toilet paper that people bought during 2020 may have helped them sleep better, but it tied up working capital that could have been invested elsewhere.

Massive amounts of toilet paper in warehouses or closets around the nation do not earn a financial return. In fact, that inventory loses value over time due to inventory carrying costs, and the potential for theft, damage, or obsolescence. Nevertheless, just-in-case inventory could proliferate at all steps in the supply chain.

This means that consumers will ultimately end up paying for elevated inventory levels through inflation, while companies suffer from reduced capital efficiency and diminished profits. In a rush to ditch just-in-time, we also risk making the bullwhip effects and impacts even more pronounced.

One simple way to address the changing dynamics ahead would be to reduce lot sizes but increase frequency of orders. This would allow for more data and greater transparency in the supply-demand balance of goods moving across the supply chain. For example, a lot size could be changed from one order of 500 per month to orders of 125 every week. That way, flexibility and data transparency would be enhanced. However, increasing order frequency may also lead to higher transportation and replenishment costs.

Prolonged issues in access to goods could lead to a protracted overhang in just-in-case preferences across end consumers — both retail and industrial. This means that in 2022, the focus on flexibility — whether in the use of automation or in levels of inventories — will be critical. ~

Daniel Stanton is a supply chain entrepreneur, the best-selling author of *Supply Chain Management For Dummies*, and a popular LinkedIn Learning instructor. Teaching under the moniker Mr. Supply Chain®, Daniel also holds a master's degree from MIT in Engineering, Logistics, and Supply Chain Management. He is currently a DBA candidate at Cranfield School of Management.

Jason Schenker FLTA® is Chairman of The Futurist Institute.

THE EMERGENCE OF PROVENANCE FOR THE CIRCULAR ECONOMY

Mark Manning

- CEO of iTRACE Technologies, Inc. -

The Origins of Provenance

Looking at the year ahead, I can sum up my expectations for the big idea of 2022 in one word: provenance. Provenance is a term that's been used in the art world for a long time to describe the ownership and history of artwork. The "Mona Lisa," for example, has a provenance. That includes its creation, its entire history of ownership, where it's been, and the fact that it's currently on display at the Louvre in Paris.

Looking ahead to 2022, both the concept of provenance and the use of the term provenance are going to be applied to things outside of the art world. It may sound surprising to be talking about very old works of art when thinking about 2022, but non-fungible tokens, or NFTs as most people know them upended the art world in 2021.

Looking ahead to 2022, NFTs may also present an important use case in helping to secure the provenance of things other than digital art. In general, a good's history, traceability, and records have been increasing in importance for years. There are a few key products for which provenance is poised to become particularly important, including farm-to-table produce, expensive items, aircraft parts, and other items with health and safety implications.

Provenance describes the story and history of something and in doing so, adds value to it. After all, an item without records that lacks provenance — particularly in the art world — has less value, as its authenticity can become questionable. For example, your used car has a provenance. It might have flood damage from a hurricane in New Orleans or it may have had one owner in the desert in Arizona. That's its provenance. If we don't know its history, it becomes a riskier and thereby less valuable item.

In 2022, the idea of provenance and the value provenance adds is likely to be more widely known and it is poised to be adopted more broadly for corporate uses across the economy.

Corporate Provenance

Provenance has always been the word people used when speaking about identifying genuine versus counterfeit artwork. We are now going to see this be applied to other kinds of assets and goods, including the provenance of parts. This is intertwined with supply chain and the security of supply chain. Furthermore, the worldwide sustainability imperative and move toward a circular economy will make provenance indispensable for vehicle and equipment parts.

Rules are changing in Europe and the United States that will foster more advances in the adoption of a circular economy and the need for provenance. In the United States, there are copyright laws in place to prevent people from tampering, changing, modifying, or even repairing certain products. This impacts software-enabled products. This description includes a broad array of goods from mobile phones to Tesla cars and medical devices. Copyright laws prevent people from getting the tools, parts, and information to repair these parts. But trademark rules have recently relaxed in the United States to ensure people have access to documentation, manuals, information, and parts to facilitate repairs.

You may not have heard about provenance in 2021, but in 2022, you will be hearing all about it. This will be everywhere, and it will greatly change the engineering of products. Devices will need to be openable, parts will need to be replaceable, and devices will need to be repairable. This will require and necessitate a change in the mindset of engineering teams toward repairability.

The "right to repair" movement started a few years ago as a grass -roots movement and is now becoming quite a force. Laws in Europe are ensuring parts and products are being made so software-enabled devices can be repaired and function. This all comes back to the environmental impact of the disposable electronic era and the circular economy of the future. Companies previously had taken the stance: if it breaks, consumers need to replace it, so I make more money. But that is changing with the push toward sustainability and a more circular economy.

Although good for maximizing profit, preventing repairs of electronic-enabled devices shortens the lifelong experience of a product. Imagine if you had to dispose of your car if the headlight went out. This has to come to an end. It has to stop. And it will.

We just can't continue living in a world full of this much scrap and waste. The whole philosophy of the one-time use component is going the way of the dodo. In the longer run, the circular economy is likely to be augmented by 3D printing of replacement parts.

The increased focus being placed on long-term value for the client, the economy, and the planet is driving companies to rethink the design of their products. This is a big transition for businesses.

For businesses and the economy, and the environmental focus we have now, this transition will become essential. Single-use electronics are ridiculous. Looking ahead, they'll be sold, repaired, and refurbished multiple times. As goods are sold, have multiple owners, and move through the circular economy, it will increase their need for provenance.

An asset moving through the circular economy will have value, but the provenance — or lack thereof — can change the value of the item itself. If one item had a full-service record and a full lifecycle, that's more valuable than the item without a history, that came out of the blue and you don't know how it's been handled or maintained or where it has been during its life so far.

Transparency of asset quality and history will be more critical for much longer-lived assets, something that can be critically informed by provenance. After all, without provenance, how do you know what's happened in the asset's history, how it has been maintained, and what condition it is in. This is essential for the aircraft and aerospace parts industry in particular.

The expected lifespan of aircraft parts is between ten and twenty years, and each part may be installed in four to five planes in that part's lifespan.

If you've ever sat at the gate on a plane waiting for a repair part, you've been exposed to these dynamics. When an aircraft needs a part, what's likely happening is that someone has to get the part from a hangar or storage facility. That part has likely been taken from another plane, repaired, refurbished, certified, etc., and then that used part gets installed on your plane before it backs away from the gate. Installing a certified used part on your plane is everyday business in the life of an airline today.

There are many questions of provenance when it comes to aircraft parts, including:

- Where was it manufactured?
- When was it manufactured?
- What airline has that part?
- When was the part last maintained?
- How was it maintained?
- What aircraft has it been on?
- When was it repaired or refurbished?
- Is it beyond its useful life?

An important thing to note about this process is that right now, the state of the art for aircraft parts provenance is a hanging paper ticket, with carbon copy paper. It's email CC in real life. As someone who travels a lot, I find this personally terrifying. We can do better. With NFTs and blockchain, I believe we will.

Supporting Provenance with NFTs and Blockchain

There are four key requirements when it comes to provenance: First, the item must be uniquely identifiable. Second, its provenance must be recorded. Third, the provenance must be permanent and unchangeable. Fourth, the provenance, the item, and the link between them must be secure and verifiable.

Linking a physical asset with a digital twin through an NFT, for example, the record of which is stored on a blockchain, can fulfill these four requirements, and secure an item's provenance. This has been the focus of my work at iTRACE Technologies.

Using NFTs tied securely to a physical item is a way to make items uniquely identifiable, which is effective since each NFT is unique. Meanwhile, a blockchain can serve a most useful function as an immutable repository for the history and provenance of an item. This works, because no one can change the history of an items' provenance as it rests on an unchangeable blockchain.

The provenance can be used to record the supply chain delivering products to retailers, and it can also be used for post-sale collecting and recording those attributes over the life of the product as it moves through the circular economy.

What you also need is a physical way to tie the physical item to its digital record as a digital twin, this is essential so you can identify the original item versus a counterfeit or fake item. You need an authentication technology that allows the item to be validated throughout its entire life. There are various open-source technologies for authentication that can be easily copied and duplicated. But the next developments of provenance require something that is a more secure link between the blockchain and the physical item than QR-Codes and barcodes.

The focus of NFTs today is around digital art. When you create a digital asset tied to a physical asset that is recorded on a blockchain, you have created a digital twin. The physical and digital assets don't need to lead separate lives. The physical item and the NFT can validate each other. That is part of the physical item's provenance. They go together and add value to each other. Provenance is exceptionally important for aircraft and other critical safety parts, very expensive assets, and assets with health and safety implications. Among expensive assets that we expect to see secured with NFTs are education and human capital assets like certifications and degrees. We expect this will be one of many advances on the horizon of the provenance discourse in 2022.

Robotics and automation are enabling the easy, unique identification of items in production giving them the born-on date that starts the whole process of creating provenance for the item. Once an item is uniquely identified, its provenance can be recorded throughout is life to ensure that it is disposed of properly with the least impact on the environment and economy. ~

Mark Manning has spent many years in the brand-protection industry, and he is now turning his attention to securing the provenance of physical products with blockchain and NFTs to help foster the circular economy, add value, and enhance consumer safety. He is the CEO and Founder of iTRACE Technologies as well as the Co-Founder of NFT Degree.

BUILDING A NEW 3D WORLD

Shane Istre

- CEO of Digital Twin Studios -

The year ahead is likely to be the year of the metaverse. We are finally starting to see the last decade of disparate technologies come together, and it will do so in the metaverse. In some ways, the metaverse is the Internet 2.0, and it will begin to morph broadly from 2D to 3D. The Internet is just a web and connection of companies, private company storefronts, and government websites. But it's 2D. It's very flat. In the year ahead, the online world will take on a data-rich 3D shape.

The Coming 3D VR World

There will be a push for creating 3D user interfaces in the year ahead. It will be important for B2C as well as B2B activities. This will be the metaverse in its practical corporate form, and we will all need to build content.

With the big Facebook Meta rebrand, we are likely to see more company realizations that they need to get on board with VR technologies. Just like in 1995 and 1996, when companies realized that they needed a website to be a legitimate business, 2022 is likely to be a year when more businesses and entities realize that they need 3D storefronts and 3D businesses to connect with customers.

Right now, Amazon has a pilot program. You can walk around your favorite retail access with an AI bot assisting you virtually. There is also AR content for houses, where you can display furniture that you don't own yet. People are familiar with that. They are familiar with 3D tours of homes they are going to purchase. A ton of houses have 3D tours but displayed on a 2D screen. More comprehensive VR and 3D offerings will emerge in the year ahead.

The Internet is ready to add these 3D assets, because all web browsers already added VR capability several years ago. This means that if you have VR content and a headset, you are ready to go. Now, once the switch flips, it's only a matter of time.

You can be helped by an AI assistant and walk around clothes in a virtual environment. This has the potential for the VR experience to include a mannequin shaped just like you based on the body scan of a camera. They are already doing this now for custom fit clothes. With mannequins all shaped like you, it is one store personalized exactly for your own preferences. It's all in one location and perfectly created just for you.

This all begs the question, why would you ever go back to 2D shopping? In truth, many may not.

The Many Metaverses

Even though Facebook just changed its name to Meta, there won't just be a singular Facebook metaverse. It will look like Facebook, but it will not be the only social media site. It won't be the only metaverse. Facebook's Meta will just be one place you visit. It's like Walt Disney World: You won't live there, it's somewhere you will go.

The amount of data in the metaverse will greatly exceed the data on the Internet. If you are alive today, you will never be able to consume 1-millionth of what's out there. Now imagine the world going from a plane to a cube. The data will be infinitely larger. AI is already working on this. Now imagine that level of data cubed. More data, more relevant, more disparate. Like the individualized nature of music apps like Pandora, Spotify, and others, we will need this exact sort of help to filter through the data and customize the preferences for your job, life, mental wellness, and more. We will need it all.

Education

In the year ahead, we are going to see VR education proliferate as well. Along with companies, universities and academic institutions will need to move into the metaverse to market themselves to would-be students. Plus, as with the transition in learning technologies from slide and video courses to 3D training like Digital Twin Studios, hardware and software have caught up, making instruction more commonly available in 3D.

Using VR to Train Robots and Self-Driving Vehicles

The virtual world presents the potential to accelerate the development of self-driving, automated vehicles. After all, 3D virtual environments can be used to train robots.

While technology integration between robots and systems used to exist in the real world, you can now integrate and train up a virtual system in the digital world. If you train AI in the metaverse, you don't need to worry about the expense and risks of the real world. This can be done across all industries, including with self-driving vehicles.

There are trucks on the road daily and automated Waymos out there right now, but by creating a virtual world of roads, it may be possible to greatly accelerate the training of self-driving vehicle systems. This works because the system does not know if the road is real or not. It's all the same to the operating system. Plus, you only have to successfully automate a self-driving vehicle or robot once to make it effective. You get scale immediately.

Costs and Headsets

VR headsets are still expensive, but companies will find ways to remove friction. They will likely become free eventually, because you'll need them to use websites. As Facebook wants you to see things in 3D, there is a rising chance for headsets to be free. Just look at the costs of Pico or Quest headsets. The cost to make them is so low now that prices have fallen, because there is nothing revolutionary with headsets.

The race to the bottom on costs of headsets could lead to subscription services. Headsets as a cost friction is likely to go away in 2022. The final piece is 5G infrastructure, and it's already there.

The Arrival of the Metaverse

Everyone is bad at estimating how quickly technology changes are coming. Five years ago, I had to explain VR. Now, everyone knows what it is. A lot can change in five years in the world of technology.

The push toward the metaverse and an immersive VR world by some of the biggest companies in the world is likely to be swift. This is all within reach right now. It's up to brands coming to that realization and coming together. The tech all exists now, and in 2022, it will begin to be everywhere. ~

Shane Istre is the founder and CEO of Digital Twin Studios.

THE FUTURE OF 3D-PRINTED CONSTRUCTION

Kaleb Steinhauer

- CEO of Genesis Dimensions -

More 3D-Printing Construction Projects Ahead

In the year ahead, I expect more 3D-printed construction projects will be announced, started, and completed. This means that 2022 will be like a football game where the ball is moving down the field at a very slow pace and you're fighting for every yard.

But the ball will still move down the field and robotics and construction of 3D-printed buildings will still move forward in 2022.

As we think about the future of the industry, there are two core dynamics at play right now for the future of 3D-printed structures. On one hand, we are in a beta phase, and printed construction hasn't reached scale.

The other dynamic at play is that there is a need for institutions around housing to be there to support the opportunities presented by 3D printing, including finance for mortgages, homeowners and related insurance, appraisers, tax assessors, and so much more.

The Value Proposition of 3D-Printed Construction

Part of the reason the 3D-printed construction ball will continue moving down the field in 2022 is because the long-term wealth creation, waste reduction, and value potential are massive. To take a step back and put things in perspective, global construction is the largest industry in the world. It is a business of over $17 trillion per year. Because of the scale of the industry, we are also not going to do everything at once. In fact, right now, if we would 3D print just warehouses in America, we would easily be the largest construction company in the world.

So, what is the market valuation going to be for 3D printing construction companies? It won't need to be a big percentage to be a massive company. For example, if we get 3% of construction worldwide, we would be not just the biggest construction company in the world, would easily be the biggest company in the world. Our material is better than what is used in construction applications today. So, our products will be even safer.

Beta Building

More and more 3D construction projects are being completed, which is good. We are still far away from mainstream acceptance. That's not happening in 2022.

You're not going to go to DR Horton in September 2022 and ask for your home to be printed. Printed structures will still be in beta for the next couple of years. The technology overall is still not up to snuff. We are not at the point where we can have a house printed for $10,000.

The beta projects going on right now are important to get the technology moving. But you might not yet be able to say that 3D printing is better, faster, and cheaper. That will happen in the following projects in the years to come, and it could quickly change the dynamics in terms of production. It will completely shatter what's out there right now. In due time, we will be able to say, "Do you want a house for half the cost in a day, at way better quality?"

But there are real hurdles. For example, in the UK, the banks are not willing to give mortgages for 3D-printed houses. But you need financial institutions to change their mind and adapt. It won't happen overnight, and you will probably have these issues and challenges for years before this is accepted as a general technology.

Robots and the Speed of Construction

Construction by its nature can be slower to adapt to new technologies. You don't have a controlled environment in construction. When you automate within the four walls of a structure, it's much more controlled.

Robots with more people makes it more difficult, because of safety. Robots in manufacturing are in cages. But in construction, it's all outside. You can't control that. You need to be able to adjust for that. It is a lot harder to adjust for that.

How do you make an uncontrolled environment into a controlled environment? If you were to create an exostructure, it would take more time to put up and break down than to build the structure itself. The solution is that you need to have a robotics system that is capable of being exposed to the environment. You cannot control the environment, but you can control your system within the environment. Everything in the robotics world comes back to rule #1: the robot needs a reference point, and it needs to know where it is in space. From that point it can operate. But the robot always needs to know where it is in space.

So, what do we expect will be the secret sauce of success in 3D-printed construction? For us, we believe it will be the ability for our robotics system to be completely mobile. Mobility is the top factor when it comes to construction. Second, our robotics system can locate itself in space, which ties back to that #1 rule for robots. Third, to achieve success, we believe the printed building material needs to adjust for external factors. There may be other recipes that can lead to success, but those are the three ingredients we are focused on.

Labor Market Issues

The labor market has gotten very tight over the past year, and the situation is not changing.

It might get worse, but that's about it. It's not going to have any improvements any time soon. For construction, those are in-demand jobs with a blue-collar aspect. It is hard work, and the labor shortage for in-demand jobs that are hard physical work is only going to get worse. But that all adds to the value proposition for the future of 3D-printed construction. It's why I am not worried about higher Fed rates, and I am not worried about a tight labor force. My only concerns about the future of 3D-printed construction are regulations. That's the only thing holding us back.

The Future Beyond 2022

We've been hearing about electric cars for decades. But they really haven't started to come into the mainstream until the past five years. For 3D-printed construction, it will be years until this is mainstream. It's probably five to ten years away.

What year is this like for printing structures (compared to EVs)? 2005 is probably a good comparison. It shouldn't take sixteen years to get to where EVs are today, but that's the problem with construction. There are so many regulations and rules. Trials take time. A good comparison is to drug trials and the FDA. You have all the regulations, which you need because it can kill people. That's the same thing with construction.

This isn't like an app or software. No one is going to die if your social app doesn't work. But we are building buildings, so people could die. There's a high threshold of regulatory proof to get that seal of approval. We will get there, but it will take time.

The most important thing when we consider regulation hurdles is that over 90% of all buildings in America are one story. That makes it a lot safer to construct with 3D printers than a skyscraper.

People may wonder, are we going to be designing a 3D-printed skyscraper? I would counter with the question, why would we? Will we do it one day? Absolutely. Talk to me in like twenty years. For the year ahead, the talk and focus will be about getting beta projects done and pushing regulations forward. ~

Kaleb Steinhauer is the CEO of Genesis Dimensions. Kaleb's solution combines large-scale 3D printing and robotics to revolutionize the construction industry through lower labor costs and limited environmental impact.

ARTIFICIAL INTELLIGENCE 2022 TRENDS AND BEYOND

Djamila Amimer

- CEO and Founder, Mind Senses Global -

In 2022, the adoption of AI will continue to accelerate across markets and economies. The combination of AI, Internet of Things, automation, and 5G will offer significant business opportunities in several sectors such as healthcare, supply chain, and retail.

The business focus will shift from pilot projects to deployment and scaling up AI applications. Key trends and applications for 2022 and beyond are as follows:

AI and Automation

The combination of AI and Robotic Process Automation enables rapid end-to-end business process automation and accelerates digital transformation. Across a wide range of industries, intelligent automation will be a key business driver.

COVID-19 has shifted automation focus from back-office functions to back-office roles in areas such as marketing and supply chain. Artificial intelligence will play a critical role in improving performance in these areas such as by predicting supply chain bottlenecks and understanding customers' personas. Customer service is another area where AI and automation can deliver significant value to businesses.

AI, IoT, and 5G

Artificial intelligence will thrive in a connected world giving rise to more business use cases than ever before. AI applications for IoT will enable companies to reduce costs and improve operational efficiencies, examples include predicting equipment failures and determining optimal operations conditions. When combining AI, IoT, and drone technology, more applications will emerge such as reducing risk and improving safety by conducting remote inspections and monitoring.

5G networks are complex and integrating artificial intelligence will address the complexity issue. The AI integration will reduce capital expenditures, improve network planning, and optimise network performance. AI will make networks more efficient by enabling dynamic network slicing and application-based traffic steering across networks.

Conversational AI

Conversational AI has gained momentum over the last three years with a variety of use cases. Thankfully, it has grown beyond simple chatbots to more complex tasks such as summarising text, sentiment analysis, and speech recognition.

The conversational AI market is already crowded with thousands of vendors and it is expected to continue to grow. COVID-19 has accelerated the adoption rate and made the market more competitive.

Customers will be looking for platforms that have advanced functionalities and which can support multi-use cases. Vendors will look to reduce their models training data needs and will focus on building multi-domain libraries. High customer expectation means that vendors need to prove that their product fulfills its promises and return on investment.

AI and Cybersecurity

Cyberattacks are growing in both volume and complexity. In the past, artificial intelligence has been seen as a threat, hackers could potentially use AI to perform a cyber-attack by enabling smart devices to be hacked or mimicking a human being at the keyboard by cracking Captcha tests.

However, cyber-security experts will increasingly start to look at AI as part of the solution rather than a threat. The power of AI could be used to analyse big data, detect patterns, and reduce response times. Artificial intelligence will be used to make systems safer. Examples include: using AI to identify threats early by spotting abnormal activities or potential vulnerabilities and scrutinising millions of websites to identify malware. The combination of artificial intelligence and blockchain will open more opportunities to address threats to safety and security in the context of cybersecurity.

Low Code AI

While business AI adoption rate is increasing there are still barriers to enter the AI market. One of the biggest barriers is lack of AI capabilities and talent. This is an accentuated issue for small and medium companies.

Low code AI will make AI accessible to everyone. Companies of every size will be able to build applications at lower costs and faster. Custom applications will be built and scaled via third party APIs. Low code platforms will be seen as the gateway to technology innovation.

Responsible and Trustworthy AI

The AI story is not always glorious. There are several examples where AI applications have failed in one way or another. Failure could be due to several factors including data bias, lack of ethics principles and a lack of human centered approach.

Examples of AI failures include:

- Microsoft's bot "Tay" that started to come up with racist remarks within twenty-four hours of its interaction with people.
- Amazon's recruitment tool, which favoured men over women for technical roles.
- U.S. Court Case Management System COMPAS algorithm, which was attributing higher risk of recidivism to black defendants compared to white defendants.
- Tech companies banning police from using facial recognition software.

For AI to be trusted, we need a trustworthy framework that encompasses the following:

- Reduction and elimination of discriminatory bias
- Testing the robustness and reliability performance of the AI model
- Understanding how AI derives the outcomes and how the data is used
- Setting clear responsibilities and accountabilities
- Complying with data regulations, safety and security

Green AI

The emergency of climate change is making governments and companies look into technology and in particular artificial intelligence to help mitigate the problem.

AI capabilities could be used in several ways such as: improving building efficiencies, monitoring deforestation, balancing supply and demand of renewable energy, and assisting in natural disaster responses.

However, using AI is a double edged sword, some of the AI models are too big and the processing power to train them can be significant. One example is OpenAI's GPT3, that has shown impressive results for several tasks such as NLP questions & answering and generating code. However, the model is a very large network composed of ninety-six layers and 175 billion parameters. It is believed that the carbon footprint for its training was equivalent to driving a car to the moon and back.

With the increased deployment of AI across economies there is a need to address the carbon footprint and sustainable aspects of these models. Focusing on efficiency of computations will become an important theme going forward. ~

Dr. Djamila Amimer founded Mind Senses Global with a mission to help businesses and organizations apply artificial intelligence, and unlock its potential.

LIFTOFF FOR 21ST CENTURY EDUCATION

Jason Gouw

- CEO of Cubit Education -

In 2020, the COVID-19 pandemic flipped the entire K-12 education system on its head. Schools scrambled to figure out what was going on, what to do, and how to do it. The most challenging problems were the questions we didn't even know to ask; the coming challenges we didn't know would come to fruition.

One thing was sure — the pandemic catapulted the education industry forward by making technology an essential platform for all learning in K-12. Computers, webcams, and the internet were leveraged by every school like never before — forcing even the most computer-averse teachers to become proficient if they were to have any hopes of continuing teaching during the pandemic.

At the start of the pandemic, the first step in establishing a viable virtual-classroom environment was through existing video conferencing platforms such as Google and Zoom.

This was a quick and dirty way to get something up and running.

Although that allowed classrooms to make a relatively quick transition from in-classroom communication to remote-classroom communication, there was much more that needed to be done for virtual classroom success; more than just voice and video between teacher and student. Some of the major challenges that needed to be addressed early into the pandemic was a lack of device and/or Internet access by students at home.

Not every student had access to a computer and not every household had Internet access; the ones that did barely had enough bandwidth to host video calls for two kids and two parents all working the network at the same time. It took some time, a bit of funding, and a lot of computer chips to pump-out laptops in order for every student to have a device.

Large Internet service providers did their part by giving free Internet access to households that were unable to afford it. In a matter of months, the foundational infrastructure was paved — something on which K-12 education can start building up from to continue the transition into a virtual classroom environment. The real challenges of virtual-learning starts here.

In contrast, most adults in the professional field are accustomed to work that utilizes digital devices for communication, collaboration, and workflow. After all, companies within the last few decades had to adopt digital technologies to remain competitive by keeping workflow and team collaboration efficient; cost-effective.

Such a competitive edge for the sake of survival was never the case with K-12 education. Many schools, especially public schools, utilized traditional methods of teaching and learning because it "still worked" and there was no real risk of "losing customers" — something that ordinary businesses face on a regular basis. There was also no incentive to drastically upgrade the entire education system overnight. This is where the old adage rings true, "If it's not broken, don't fix it." Students and parents are the customers of public schools and by law, kids are required to go through K-12 schooling. In other words, schools have an endless supply of customers year over year, customers mandated by the government to attend these schools. Coupled with the fact that the education industry is a gigantic, slow-moving ship with lots of complex parts, rusty paint, and overdue maintenance items, having the latest and greatest technologies in a school is only a luxury few districts in tech-centric areas manage to implement.

It's funny that sometimes it takes running into a wall or dead-end for us to try something dimensionally new. It causes us to stop looking left or right for a way out and instead, to consider looking up to find a fundamentally new way to evolve.

Once the infrastructure was established to allow teachers the ability to remotely teach their students, K-12 classrooms found it necessary to have at least two teachers per class. One teacher would be the lead, focusing on delivering the instructional lessons while the other managed the video conferencing platform and addressed students' "hand raise" in the virtual rooms. Having two teachers per class was certainly not sustainable or even doable by the majority of schools — even with classrooms doubling up to have two teachers working together at the same time, which also meant twice the number of students to manage virtually.

Deeper challenges in a virtual setting involved things that were beyond a teacher's control. Students that kept their cameras off, possibly dozing back asleep, students that played games while logged in the virtual class, or young students that simply didn't know better, became bored, and closed their laptops — these are things that could not be addressed by the teacher. The quick workaround was to get the parents (working from home) involved. They would serve as an emergency backup, a standby for accountability to ensure that a student logged-in on-time and had their video and audio working.

This, again, only worked with a small portion of the student body; and the parents that were able to participate felt the burden as they, too, are trying to get work done from home without too many interruptions. Among all of these challenges encountered during the transition to virtual-learning, one major challenge is, for the most part, unresolved.

In STEM subjects, this one challenge, perhaps, poses the greatest loss to a child's education — the inability to experience and gain skills through hands-on experimentation — a critical skillset in the increasingly important STEM fields.

Workarounds using simulations provided a theoretical sense of the experience but failed to teach the critical observational and troubleshooting skills associated with hands-on projects. This was a big issue in STEM since part of the required science standard, for example, was to learn lab proficiency and safety processes. How can a teacher mark a student proficient if they've never actually learned to operate microscopes, micrometers, or sensitive scientific instruments with their piers?

Plausible Solutions

Getting thrown into the fire after being in the frying pan isn't always a bad thing. New growth and opportunities emerge in a hot crucible. The education industry benefited from the pandemic in that all of its teachers became proficient with computers.

For some districts, it used to be pulling teeth to get teachers to start using computers and software like, Google Classroom, for example. Now, just about every teacher can log into a computer and navigate their way through various programs and platforms — essentially new tools that they've become comfortable using. Just as beneficial is the fact that almost every student now has access to a computer and the Internet at their home — allowing them to join classes virtually and so much more.

They can now research new topics on the internet at any time — learn things that they're curious about — something many families didn't have the luxury to afford.

Though we all hope to get our kids permanently back into the classroom and never have to immerse them again in virtual learning environments, new technologies have been developed as a response to improve the virtual learning experience. One of the biggest things brewing in education technology is the use of AI seamlessly in the background of virtual classrooms to filter out unwanted noise, adjust camera settings for picture clarity, normalize voice volume levels, keep tabs on student activity and engagement, and much more.

Looking at NASA's 2021 Space Apps challenge — the majority of the challenges created by NASA tasked students to find ways to optimize the organization and collaboration of remote teams working together on a project through clever software. Unlike a regular Zoom call, these remote collaboration technologies seek to automate data sharing, version control, and remote collaboration of hands-on projects. This time, it's getting scaled up to very large group sizes (think crowdsource) and catered for more complex tasks instead of the weekly Monday morning team meeting.

This type of remote-collaboration technology could be a paradigm shift to how teams have been bundled locally to work on projects — especially projects with physical, hands-on attributes.

Projects would have the opportunity to tap into various pools of talent from around the world, not just in local areas. I see AI being used to blend the human-machine interaction to make the inter-human connections through cyberspace as natural as a group of scientists and engineers assembling a prototype robot in the same room. The course of the next year is still in uncharted and wavy waters but the technology is here and the possibilities are only limited by the constraints of our imagination. What an exciting time to be alive in the world of technology. ~

Jason Gouw is a learner of all things and founding CEO of Cubit Education. Jason's eclectic interests span engineering and lumberjacking.

CREATIVITY AND COMPUTATION

Louie Kangeter

- President and Co-Founder of Fanbasis -

The year ahead will bring some important changes for how technology impacts the creative economy. At its essence, the creative economy can be defined both by the production of content, art, and entertainment, as well as by the value creativity brings to the wider market in high level problem solving, executive functioning, and innovation.

Thanks to increasing capabilities of AI and automation, we will see the creative economy surge towards community ownership, and gamification. At the same time, automation will fulfill more roles than ever before, and people will be incentivized to join virtual communities by large corporations hoping to cash in on the metaverse. We expect to see individuals seeking to find new ways to monetize their talents, and invest in communities, especially as traditional jobs become more and more difficult to acquire.

As automation fulfills more functional roles in society, creativity and entertainment will become the currency of choice — difficult to automate, yet increasingly easy to own. Most experts agree that programming creativity is one of the great challenges of artificial intelligence, and as automation becomes more prevalent in society, it is clear that people will seek human connection in their daily lives in new venues. Digital spaces that are currently structured as one-way streams of entertainment from creators to consumers will be forced to adapt to an age-old desire — the desire to connect. How does this materialize today? Market leaders in the creator space are focused intently on building platforms and infrastructure that empower creatives to provide experiences and valuable content directly to their fans. The widespread adoption of technology like tokenization will provide consumers with the ability to participate and invest in their favorite creators, developing community based economies around creative individuals.

Most importantly for creators, fans will be incentivized to invest early in content and creators that they support or find value in, allowing more individuals to experiment with monetizing their creative skill sets that are the most defendable bastion from widespread automation.

The creative problem solving space also holds intense potential, especially in the realm of gamification. Gamification of problem solving is not a new concept. It's brilliantly encapsulated by Orson Scott Card in the 1985 novel *Ender's Game*, which describes the rise of a ruthless and creative military leader through simulations that are eventually revealed to be real world conflicts.

As micropayment infrastructure and game tokens become more and more accessible to game developers, players will be rewarded with currency that is not only usable within the games they are playing, but broader ecosystems that can contribute to the traditional economy. With the rapid development of metaverse technology, players will soon be incentivized by companies like Roblox, Meta, and Disney to take full advantage of new technology that comes with financial benefit. Using games to simulate real world situations, then pulling the behavioral data from those games and applying them to AI as learning tools provides spectacular promise in the ability of algorithms to understand human behavior. Finding ties between gaming behavior and real world interactions is a key to the creative economy contributing to, and not competing with, automation and AI.

The rapid growth of AI and automation technologies has some industries excited, and some labor forces scared. Just like the rest of human history, this is merely an evolution in tools that will shift behavior, not the beginning of a skynet-esque end of humanity. The future of work will change, and the definition of what labor is will change, as automated processes handle more and more tasks and leave humans to what defines us - creativity and innovative thought. ~

Louie Kangeter is the President and Co-Founder of Fanbasis, an influencer platform that allows fans to interact in countless ways with their favorite creators, athletes, and celebrities.

BULKY, AWKWARD ITEMS: SOLVING THE NON-CONVEYABLE CHALLENGE

Bruce Bleikamp

- Director of Product at MHS -

Good news! That new exercise bike you ordered arrived. Now, the bad news: you have to move it upstairs! The flat-pack box is lean, but also long and wide — so much so, that it barely fits up the stairs.

What you're experiencing in the struggle of moving such a bulky, irregularly shaped item through your house is reminiscent of the challenges that distribution centers face handling such items every day.

Whether furniture, sporting goods, outdoor equipment or something else, large, irregularly shaped items are described as "non-conveyable" when their physical characteristics prevent them from being handled by automated conveyor and sortation systems, necessitating alternative workflows.

For example, the length of an item, like a golf club, can cause it to get stuck on an incline. The inconsistent center of gravity of a round item, like a bowling ball or packaging tube, can prevent it from starting and stopping on a conveyor. Last but not least, the size and weight of items like furniture or exercise equipment can simply exceed the dimensions compatible with existing automation.

Handling such items is not a new challenge. However, growth in online ordering and direct-to-consumer models compounded by the effects of the COVID-19 pandemic have caused a surge in the amount of these large, awkward non-conveyable items. Not only that, responsibility for moving such items and delivery frequency has also changed.

In the past, rather than shipping directly to consumers, logistics operations moved large items like appliances and furniture for retail replenishment. Retailers then took the baton with their own established last-mile delivery processes, or consumers simply hauled items home themselves.

The expansion of e-commerce, which is expected to gross $4.2 trillion globally in 2021, shifts the burden of handling such non-conveyable items. Businesses solely dedicated to online ordering and direct-to-consumer delivery of products like furniture and outdoor equipment are creating logistics operations wholly focused on handling large, irregular items. But the impact does not stop there — the downstream effect of greater non-conveyable volumes affects parcel carriers, too.

This trend was underway before the COVID-19 pandemic began, but its lasting impacts have only increased the number of non-conveyables that need to be handled, with consumers shifting even more of their purchasing away from brick and mortar, as evidenced by the 64% of people who have purchased oversized goods online.

Labor challenges only compound the difficulty operations face, as they process more bulky items with fewer resources to handle them.

Under these conditions, more businesses stand to benefit from new automation solutions designed to handle the growing niche of non-conveyable items.

Disadvantages of Manual Handling

From an operational standpoint, facilities take steps to consolidate items to more efficiently manage material flow throughout the facility and keep the dock organized. But how do you handle large, irregular parcels?

The show must go on, even with non-conveyable items, so operations employ alternative means to take care of the consolidation process, such as loading those parcels onto carts. But no matter the workflow, moving large, bulky items through the distribution and fulfillment process comes with major drawbacks.

Transporting non-conveyables from point to point can lead to lots of lift truck or tugger traffic, bringing safety risks, delays, and ultimately congestion and disorganization on the dock. It also uses a significant amount of labor, with employees required to manually lift and awkwardly twist in ways that can lead to repetitive stress injuries.

The U.S. Bureau of Labor Statistics reports nearly 150,000 back injuries per year in the workplace — and the business impact of each back strain adds up quickly. Direct costs attributed to medical and legal costs — as well as OSHA fines — can add up to thousands of dollars for each injury, though the amount an employer pays depends on its workers' comp policy. But employers are always on the hook for indirect costs, which can also reach tens of thousands of dollars through lost productivity and damage to equipment and facility infrastructure.

Productivity is diminished both by the time that the employee is out of work as well as the less-efficient performance of a new employee who lacks the same level of training and experience as the employee they replace. In total, OSHA estimates businesses spend $170 billion per year on costs associated with occupational injuries and illnesses.

Not only that, the transient nature of the labor pool makes it difficult to rely on manual processes. Turnover keeps getting higher in the warehousing industry, rising from just over 40% in 2016 to nearly 60% in 2020.

Evaluating Workflows for Automation

An investment in automation to handle non-conveyable items can bring a quick ROI and significant savings, including reduced labor dependency and risk of on-the-job injuries and related labor costs. Operations should take inventory of the rate of non-conveyable items to be handled, identify all the workflows involved in the handling process, and estimate the associated labor requirements.

The guiding principle is to automate the most mundane, repetitive part of the process and leverage employees to address tasks that require the unique capabilities of humans, like quality assurance inspection. For example, automate repetitive, point-to-point transportation with a low-cost automated guided vehicle (AGV) solution that avoids the risks associated with having employees maneuver loads through congested travel paths.

Customizing the Automated Solution

Because each operation is unique — with an ever-growing number of different product types, movable obstacles, and rate requirements — an automated solution requires an inherent degree of customization.

For example, a robotic arm may be necessary to load heavy items onto an AGV or a custom package of several more specialized automation pieces like right-angle transfers may be necessary for conveyor systems to accommodate long items.

Each type of non-conveyable item and warehouse environment presents its own unique challenges. Finding the right balance involves meeting requirements without overwhelming your operation with additional functionality that is both unnecessary and costly.

To make an analogy, despite what teenagers might say, they do not *need* a luxury sports car or monster truck to go to and from school every day. In fact, something more practical and affordable is perfectly capable. For the purposes of handling non-conveyable items, the imperative is to develop a solution purely based on the site-specific realities — without bias toward a particular technology or bending operational reality in order to match a specific product.

Automated MHS robot carrying packages.

Trying the Solution on for Size

A pilot program offers an important step to get a solution for non-conveyables off the ground. It provides an opportunity to not just anticipate but *answer* key questions about AMRs or AGVs before a full implementation.

At a high level, a critical consideration is the ability of automation to scale. For example, "flexible" automation like an AMR or AGV system that leverages several individual equipment assets enables you to add vehicles, change paths, and shift workstations with relative ease.

Compare that to "fixed" automation, which can require shutting down systems or even removing and reinstalling rigid infrastructure to accommodate the same level of change. Operations can expand a sorter, but that requires taking it offline. Adding more AMRs, however, does not require shutting down the rest of the fleet.

Putting Automation into Practice

Consider deployment, and in this day and age, that includes lead time. For example, some automation projects can include wait time from suppliers, engineering, installation, and commissioning — easily extending lead times to six months. Whether fixed automation or the flexible paradigm of AGVs and mobile robots, consider what options can get your operation up and running in time to meet business requirements.

Regardless of whether non-conveyables account for a facility's core business or are simply a rapidly-growing "necessary evil" they must deal with, they require a level of customization and help from a consultative resource to find the right fit. ~

Bruce Bleikamp is responsible for equipment and technology solutions within the distribution and fulfillment division of MHS. Bruce has over forty years of experience with material handling systems serving a broad range of industries.

ROBOTICS AS AN INVESTMENT IN STABILITY

Steven LaFevers

- VP of Emerging Technology at Hyster Company -

For operations throughout the supply chain, disruption and instability have become the rule rather than the exception in recent months and years.

The COVID-19 pandemic presented unique challenges, with associated lockdowns and public health effects wreaking havoc and causing delays for finely tuned supply chains. But that was just the beginning — semiconductor and commodity shortages led to temporary plant closings and a shift to a more proactive approach to component sourcing, rather than the purely lean methods that have guided manufacturers for decades.

Already difficult labor challenges were brought to a head, with total annual turnover spiking to a five-year high of 57.3%.

Because of this, just-in-time philosophies are giving way to a resurgence of "just-in-case" planning, but the underlying drive for maximum efficiency remains exceptionally intense at all levels. As operations re-examine their long-term strategies with a renewed focus on shoring up business continuity and stability without incurring undue costs, automation has emerged as a key tool.

In fact, the adoption of robotics and automation is forecast to more than double over the next six years, springing from 38% adoption today up to 83%, according to the 2021 MHI Annual Industry Report's survey of supply chain and manufacturing leaders.

Hyster Robotic Lift Trucks (RLTs).

Fortifying the Labor Force

Against the backdrop of high turnover and a severe labor shortage, improving employee satisfaction and retention is critical to limiting the costly cycle of hiring and retraining workers — exercises that can quickly swell labor expenses. But what do robotics have to do with keeping employees around and supporting a stable workforce?

Academic research shows that organizations augmented by automation are 33% more likely to be "human friendly" workplaces, in which employees are 31% more productive. And according to a Gallup study, organizations with better employee engagement achieve substantially better retention.

Robotic lift trucks provide real value by taking on the most repetitive, turnover-prone tasks in material handling operations, such as basic trash retrieval, milk runs and putaways, and allowing workers to focus on more meaningful, value-added roles.

These robotic lift trucks are not the same as traditional AGVs used in manufacturing — they can adapt to obstacles and move around them. They can also be reprogrammed to adjust routes and remain a long-term solution as business needs evolve, combining the predictability and reliability of automation with the flexibility to adapt to the reality of the shop floor and changing layouts.

Whether using robotic tow tractors to move carts into position to keep assembly lines moving or using robotic stackers to pick up palletized loads at the end of production lines, robotic lift trucks can help enable the stable, predictable material flow. But horizontal transportation is not the only capability — robotic reach trucks can store and retrieve pallets from storage locations 30 feet high, opening up a range of additional tasks for automation.

Robotic lift trucks are not a direct replacement for human operators. Not all manually operated lift trucks are available as robotic models and under most circumstances, it takes 1.3 robots to achieve the output of one operator. Real value comes in stability, automating the most turnover-prone tasks to help reduce an operation's exposure to turnover in roles that are challenging to keep filled, especially in demanding, intensive industries.

Bolstering Consistency and Productivity

In today's high-turnover labor paradigm, operations commonly depend on less-experienced employees or those without a lot of experience at that specific site. Distracted or inexperienced lift truck operators may drive too fast, cut corners too sharply, or even drive through prohibited areas, which can lead to mistakes, damage and injuries. By contrast, robots follow programmed site-specific rules of the road such as maximum speed or minimum distance from pedestrians or objects. But as supply chain difficulties and rising costs lead manufacturers to raise prices, does the investment in robotics for stability mean the cost of doing business gets even higher?

Reducing Operating Costs

Investment in automation requires considering the cost of entry, the ongoing financial commitment and several important details related to the application.

In a three-shift application, a robotic lift truck project can achieve a return on investment in 12 to 18 months, or in 18 to 24 months for a two-shift application. The exact timeline depends on several variables, including the quantity of manual units replaced, the quantity of robotic units purchased, the total operating hours, desired throughput, and the burdened labor rate.

For example, say an application with two shifts per day, operating five days per week needs to move palletized loads from the end of a production line to outbound staging or storage. This requires three full-time operators paid at a rate of $19 per hour, plus two hours of overtime each week. When four robotic counterbalanced stacker trucks are used instead, the operation sees full payback in just two years, and after breaking even on their investment, they save $246,000 per year in hourly labor costs.

Looking to the Future

If the current supply chain climate has revealed anything, it is that ROI and cost are not the only factors that guide capital planning. Supply chain disruption has many leaders choosing robotics and automation to hedge against the threat of volatility and long-term labor challenges related to availability.

As unique global events and related supply issues continue to spur change, both in manufacturing and elsewhere in the supply chain, technologies including robotics have an important role to play in helping operations adapt. ~

Steven LaFevers is Vice President of emerging technology and responsible for the growth and profitability of a comprehensive suite of technology-driven Hyster® products and services that accompany materials handling equipment investments. The emerging technology suite of solutions includes market-leading offerings in robotics and automation, motive power and telematics.

DIGITAL TRANSFORMATION IN SUPPLY CHAINS: WEIGHING BEHAVIORAL VERSUS MACHINE-BASED DECISIONS

Robert Handfield

- Professor at NC State University and Author -

The current set of supply chain disruptions is bringing increased focus to the need for digital transformation of our supply chains. Nowhere is digital transformation needed more than the Los Angeles port, where major operational disconnects are occurring between ships, trains, and trucks. Compared to other ports like Hong Kong, Shanghai, and Rotterdam, the L.A. port is long overdue for a complete redesign of its digital infrastructure.

In particular, the need for improved integration of scheduling inbound container ships with advanced planning and scheduling of inbound trucks and trains into the terminal is one reason the L.A. port is only half as productive compared to other ports around the world.

This is one area where machines can clearly do a better job of processing many complex activities, because the outcome (increasing the throughput of containers and limiting the wait time of drivers) is clearly expressed. However, not every supply chain and procurement decision fits neatly into this category. I believe that automation is appropriate for the "right" set of supply chain tasks, that do not require consideration of context, interpolation, or double entendre situations.

For instance, many supply chains find themselves in periods of increased volatility — both in the form of customer-facing demand variation, and as seen in supply-side shortages and capacity limitations that are impacting delivery of goods and services. The vibrations that occur between supply and demand can cause considerable swings, and this can seem to be a chaotic situation. To address this issue, category managers in procurement may increase the planned lead times in an MRP reorder point system to ninety days, or they may simply enter it at the face value quoted by a supplier, without further negotiation with the supplier.

This could cause inflated planning lead times to infiltrate a planning system, because in a planning system, buffer stocks and lead times are all fixed, and those determine the outcome of the balance sheet (inventory on-hand). Instead, if the category manager questioned the lead time on material, they might be able to get to fifty to seventy-five days in a negotiated settlement, which would substantially improve the responsiveness of the planning system and increase velocity. If machines were doing this work, it wouldn't question lead times.

Tom Linton notes that when he was the CSCO at Flex, he required all purchasing managers to slash their supplier lead times, as well as their planned internal assembly time. He did so on the basis of the assumption that humans were making subjective decisions on what lead times to load into the system. In this case, a human intervention can be superior to a machine-based one.

A different example can be observed when one considers the situation of N95 mask supply during COVID. Because of the outbreak in Wuhan, the lead times for masks increased overnight. It likely took weeks for this information to filter through supply chains, and suddenly several weeks later planners came to realize that N95 masks were unavailable, and that the lead times had grown to nine months! A machine would have transmitted this information instantaneously through supply chain planning systems and people could have acted much sooner to expedite or understand the problems with masks. This is a "positive vote for the machine." The point of this is that human context and decision-making has to be weighed off against instantaneous data interpretation. There is a real benefit to the social interaction of procurement, which leads to improved collaboration on difficult and complex problems. However, machines can be less judgmental, and render decisions without having to take pause, wait for their boss to come back from vacation, wait to get approval from a superior, or think about it for a few days. A new book called *Noise* by Kahneman, Sibony, and Sunstein reveals the incredible disparity in human decision-making, which can result in decisions which vary by as much as 40-75% when expert decision-makers are faced with the same set of information about a decision.

This applies to multiple areas, including insurance premium adjusters, physicians reading radiology reports, courtroom judges handing out sentences, and many others.

The point here is that machines can often make better and more consistent and less biased decisions with less noise, when humans are faced with disparate decision outcomes even when presented with guidelines. On the other hand, in the book *Army of None*, Paul Scharre shares how a machine may mistakenly identify a child or a passing dog as an enemy and kill it with a bomb strike, as it does not discern between moving objects and assumes they are always enemies if they are in the attack zone. The same logic applies to a planner who hears about lead time increases, and raises the minimum order quantity — the decision as to how much to raise it is better made by a machine, while the actual lead time negotiated may be better suited to a human decision-maker.

An important component that drives decision-making in an automated context is how you are measuring outcomes. Work activity is typically influenced by how people are measured. If you are measured on cost, you will drive all of your activity to achieve that end, and a machine will as well. If you are measured on speed and velocity and reliability, then cost may be a tradeoff that has to be made. Recognizing these tradeoffs is at the very heart of the supply chain disruption dilemmas we find ourselves in today. And it's time to face the music, as we have been too focused on cost for too long, and are paying the piper for those decisions today.

Training Supply Chain Executives for the Digital World

With the movement towards digital transformation, many procurement and supply chain executives are pushing for more technical skills, given the increased focus on IoT and more advanced Operations Research applications for SCM, including non-linear optimization, simulation, and stochastic agent-based modeling. However, we are also positing that we cannot forget the need for strategic foresight, planning, and what the futurist Peter Singer calls "Useful Fiction."[1] Useful Fiction, or "Intelligence Fiction," is a deliberate fusion of narrative's power with real-world research's utility. We are moving to a world of machine-speed convergence of possibility that often leaves conventional forecasting or analysis out of step.

So, FICINT melds communication and envisioning techniques honed by creative communities of writers or graphic novelists with the process and rigor of intelligence collection and analysis. In some circumstances, it can be an act of prediction, in others explanation.

As organizations struggle to deal with an unpredictable ecosystem, we need behavioral skills that are advanced, which include the ability to create intelligence networks to be able to both "grease" the flow of supplies and services, as well as increase our ability to detect potential risks to that flow via human sensors and intelligence gathering.

It is sometimes difficult to convince recruiters that behavioral skills are important. Today it seems that everyone is on the tech skill bandwagon, who can work directly with improved emerging technologies like blockchain, control towers, more advanced warning algorithms, etc.

What we need is a more nuanced position, that addresses the fact that behavioral skills are lacking but tend to yield greater benefits. In particular, we need experienced procurement, contract management, and supply chain managers who have the critical behavioral skills needed to deal with the "nuanced" decisions that we alluded to earlier.

The importance of behavioral skills was supported by a Harvard study (conducted in 2020 and released by The National Bureau of Economic Research)[2] that looked at what the differentiating factors were now that "most" large organizations were digitally capable. The findings were that people had become the differentiating factor — hence the importance of behavioral/ relational skills. In short, Emotional Intelligence was more important than IQ once the other types of digital tools were in place.

We also need to give some thought as to the different roles that procurement and supply chain managers play on teams. In general, we find that they can be in the following categories:

Jacks of all Trades tend to be able to adapt quickly to situations in a supporting role. These people perform best with medium-level activities. The pool of available people is typically high and varied. Digitalization is aggressively trying to automate these roles. Jack of all Trades get the critical mass of work done but sometimes miss the critical details. They are often best at managing low complexity tasks.

On the other end of the spectrum, **Experts** tend to own a single, focused area and manage the environment surrounding it. It takes a long time to develop someone as a high-performing expert. The pool of available people is typically very small. To develop someone from scratch takes many years. Experts can deal with much more complex tasks, and are usually able to take decisive action quickly.

Teams tend to own multiple products and/or services and manage the environment surrounding them. **Specialists** lead when needed and support the team at other times. Teams are fluid and able to accommodate people moving in and out. A product/service focus makes teams more tuned to customer needs. Teams have the expertise to ask the right questions and find the right people with the best answers. But they may take much longer to make decisions.

A common challenge is to understand where and how to put people into the right types of roles. As leaders look at converging supply chains, contracting, and procurement, a key issue is who should have the balance of power: the individual Specialists and Experts, or the leaders who may be the Jack of All Trades?

Getting the right people in the right roles is critical.

For instance, one executive noted that in a large global automotive manufacturer back in the late 1990s, the top engineer at the factory had a more senior role than the plant manager with responsibility for 3000 people. Each respective pathway was designated as "Management" and "Professional" (technical), implying that they were two distinct sets of capabilities. The key problem here was that top management leaders were leaving the company, because the only jobs they could be moved into were technical jobs that they either couldn't do or that weren't using their core skills.

A similar problem is that of the "unicorn" management role in SCM/procurement. We are consistently observing that large bureaucracies cannot pivot swiftly or effectively enough to manage the true unicorns in their workforce. For instance, a senior Air Force officer notes that "my predecessor was a Contracting Officer who also happened to be an expert data scientist and computer programmer (self-trained but phenomenal). He was using algorithms used by the DoD and Google. Yet we pushed him into the traditional officer path when he clearly wanted to leverage his skills for more advanced work that others couldn't do. The system simply couldn't compute that concept." This may be a problem as we start to look at the types of roles needed in procurement/contract management, and supply chain.

Most procurement people are in the cost control or compliance mindset. Some of the more advanced are lean thinkers but have no path or method for dealing with the unknown. The unicorn worker may have advanced skills for which a job description doesn't exist, but which the organization definitely needs. In such cases, companies feel the need to outsource rather than see what they have organically. These and other issues will require further research and analysis in the near future. ~

ENDNOTES

1. https://www.queensu.ca/psychology/sites/psycwww/files/uploaded_files/Graduate/OnlineJournal/Issue_2-Singer.pdf
2. https://www.nber.org/system/files/working_papers/w28733/w28733.pdf

Robert Handfield is the Executive Director of the North Carolina State University's Supply Chain Resource Cooperative. The SCRC is a university-industry partnership dedicated to advancing the supply chain industry and the professionalism of its practitioners.

SEVEN TASKS PRIMED FOR GROWTH IN AUTOMATION

Kevin Paramore

- Emerging Technology Commercialization Manager,
Yale Materials Handling Corporation -

Labor availability is a longstanding challenge for supply chains, and current data indicates operations are looking for relief. According to the 2021 MHI Annual Industry Report, more supply chain leaders rated hiring and retaining qualified workers as extremely or very challenging than any other issue. With warehouse turnover at a five-year high of nearly 60%, automation continues to emerge as the solution operations are looking for to relieve their dependency on a transient labor pool, reduce operating costs, and help increase efficiency as they face escalating demand for high throughput and fast delivery.

By shifting certain applications to robotics, companies can focus their employees on more engaging, value-added tasks that require human judgement — leaving the non-stop, tedious, back-and-forth load moving to automation. But what might this growth in automation look like on the shop floor?

For companies serious about using automation to augment an increasingly limited labor pool, the logistics processes listed below are strong candidates for robotics.

1. From Production Line to Storage or Shipping

Finished goods need to move quickly to a storage or shipping area. These jobs are usually repetitive, longer hauls that can take two minutes or more per load. When products come off a palletizer or stretch wrapping system ready to be picked up, a warehouse management system (WMS) can signal a robotic counterbalanced stacker to pick up the pallet and move via the most efficient route to its assigned destination. For heavier loads or areas that require a tighter turn radius, a robotic reach truck may be the best choice.

For loose products not on a pallet, a robotic tow tractor with a cart can be programmed to position itself at the end of a conveyor or sorting table to receive products. Once a sensor or worker indicates the cart is full, the WMS system directs the tow tractor to shuttle that load to a final destination in a storage area or a shipping lane.

2. From Receiving or Staging to Put-Away in Racking

The receiving docks at a high-volume warehouse must quickly move incoming loads to make room for the next arrivals. Once pallets are offloaded from trailers, inspected, labeled and scanned, the WMS triggers a job for a robotic reach truck to pick them up and move to a designated drop-off location in the racking.

The WMS provides the exact pick-up spot and drop-off point so the robot accurately places the load in the right place. In fact, robots are more accurate and reliable than their manual counterparts when it comes to placing loads in the correct racking slot every time. Even in facilities without a WMS, some robotic trucks have specialized features to detect pallets in pre-programmed locations and create individual missions to put the loads in storage.

3. From Raw Goods Storage to Production Line

When a manufacturing line runs low on raw goods, fast replenishment is essential to preventing idle time that puts production targets in jeopardy. For a consistent supply, robotic lift trucks (RLTs) can be programmed for automatic deliveries or buttons can be installed for workers to make a manual request when parts are low. A robotic model then picks up the necessary material from bulk storage or racking and drops them off to feed the appropriate line.

This replenishment application is a common automation target because it keeps workers on the line and in controlled areas. Fewer workers moving widely around the facility keeps employees on task and helps mitigate the risk of injury associated with high traffic areas.

4. Picking Tunnel Replenishment

In this workflow, employees concentrate on building pallet orders in picking tunnels while robotic lift trucks handle replenishment — keeping products available to pick without disruption or delay.

When a SKU runs low, the WMS triggers a request to replenish, and robotic reach trucks retrieve inventory from storage and bring it to the pick tunnel to support uninterrupted order fulfillment.

Different picking tunnels receive product differently, but some common applications for automation include depositing pallet loads onto flow racking from outside aisles, into single or double-deep racking outside the tunnel or in a bulk staging area outside. In most scenarios, the robotic truck remains outside the pick aisle to avoid impeding the manual or automated operation. If storage is located above the picking tunnel, a robotic reach truck can automatically pull pallets down from high-density vertical and double deep racking when a replenishment request is triggered.

By automating storage and replenishment work, operations can boost labor efficiency, with employees isolated from high-traffic areas and focused on the more challenging, engaging tasks of picking products and building pallets.

5. Cross Docking

Robotic lift trucks excel at performing with the speed and accuracy necessary in cross-docking applications. When a load is received and scanned into the WMS, a robotic end rider can retrieve up to four pallets at once and deposit them in the designated outbound locations.

Automation is especially valuable in this workflow because cross docking typically involves simple pallet moves. Assigning robotics to address this task allows operations to better utilize experienced operators for work that requires more complex maneuvering, rather than tying them up with basic point-to-point transport.

6. Trailer Loading and Unloading

Loading and unloading trailers is an especially repetitive and predictable task that companies can economize by using robotics in place of scarce and increasingly costly labor. Automated trailer loading vehicles are best suited to provide ROI in environments where trailers are unloaded around the clock, every day. For example, robotics can be used for continuous loading and unloading of trailers in high-velocity facilities, whereas manual processes require breaks and may work at a declining pace due to fatigue.

However, loading and unloading are more challenging applications for automation because of the unknown condition of the loads and the potential for pinwheeling pallets and dock locks that can prevent robotic lift trucks from servicing the trailer. RLTs need consistent pallet quality, stable loads and some dock equipment to ensure safe loading/unloading.

7. Receiving to Staging, Staging to Outbound

A robotic end rider or counterbalanced stacker can be deployed to continually move loads from a dock zone or staging area to a racking lane, staging area or outbound location.

These automated jobs are very simple for warehouses to implement because the lift trucks can detect pallets themselves, no full WMS integration required.

Leveraging robotics for these jobs can help operations reduce costs by making more efficient use of their labor — an especially effective cost reduction method considering labor is often a warehouse's largest expense, typically consuming anywhere from 50–70% of warehouse operating budgets.

From Automation Plans to Automation Action

Having matured past the awareness stage, robotic lift trucks are now playing a major role in helping warehouses and manufacturers meet demand despite a limited labor supply and high turnover. In fact, robotics are poised to reach 76% adoption in the next five years, according to MHI. Against the backdrop of an increasingly demanding logistics market and tight labor situation, long-term plans for automation can be expedited to keep pace with order volumes that keep getting higher. ~

Kevin Paramore is part of Yale's emerging technology team, which provides problem-solving solutions that leverage the latest technologies, available through national account and dealer sales channels. He has over fifteen years of experience leading sales teams and a bachelor's degree and MBA from East Carolina University.

VOICEBOTS ARE THE FUTURE OF EDUCATION

Lorenzo Carver

- Co-Founder of Brainsprays.com -

Meet Your Next Teachers: Voicebots

Voicebots are going to be the big thing for education in 2022. Big companies have been using voicebot technology for years, having voice assistants go over transaction documents, rather than have lawyers to do the work. But when we start talking about teaching people things, there are a lot of emotional layers that can be removed or reinforced to ensure that there are potentially better learning outcomes with a voicebots than with a real person.

In some cases, you may also have a real person there, like a parent, teacher, or tutor, and supplement that with a voice assistant. This allows for efficacy to be assessed and measured in real time. I imagine that in 2022, we are going to see this explode. This is not just a supplemental technology for early child ed or higher ed, it's going to become invaluable for every kind of education, including for workplace and professional education.

Amazon's Alexa is an excellent example of a means to implement this, and it shows the value. Perhaps its most important attribute is unlimited patience. This makes it invaluable for people trying to develop language skills. You can keep speaking until the words or phrases are recognized. Without outside interference or feedback, voicebots can ensure language skills and techniques are demonstrated correctly.

There is also a major benefit for older learners as well as for executive and professional learners, because people can speak to the bot without fear of embarrassment. This ensures and helps protect people's egos. I expect this will take over education. In the next twelve months post-COVID, people will realize there are gaping holes in education systems and that progress has been uneven. I expect people will realize that there are even bigger issues in the areas of previously undiagnosed learning weakness. This reality could make voicebots important and essential for closing learning gaps and deficiencies. People are looking to technology to close those gaps. And whether voicebots tech is married with other devices, or whether it remains audio-only, there will be tremendous opportunities to close those gaps and rectify certain learning issues that have been revealed by COVID-19.

Why Voicebots?

When you speak things, you learn them in a different way than if you internalize information. You can see this with class participation and children's abilities to become proficient. This is something tough to do in a classroom, because classrooms are not conducive to individual learning.

When you have multiple learners, you usually have different learning capabilities among any group. There may also be a not insignificant requirement for individual learners to take time to build rapport with other learners. This is not a requirement or concern when we are working with audio-assisted learning.

Today, companies have a voice actor go through an entire alphabet and lexicon, changing inflection and intonation with automation and machine learning that allows them to edit voices. This allows for individualized learner customization along with personalized learner optimization.

Many learning environments entail teaching "to the middle" where you teach to the average ability of a class. With audio-enabled education, this is irrelevant because it is designed to address the needs of individualized learners. The goal isn't just to keep up with the average, you need to have education propel people past the average. And the only way to get ahead is to push forward.

This is not something you can even get from one-on-one instruction, because tutoring hours have limits. School hours have limits. But there are no time limits for audio-enabled, customized learning built on a platform of algorithms. This also applies to education in a lot of different areas. In complex subject areas, this allows instruction to be broken down into much smaller parts in a way that ensures that learning goals have been met before moving on to the next level or next lesson. This helps provide real time visibility in terms of learning, knowledge retention, and capabilities.

On the flip side is the cost. Not only is the technology superior, but the cost is so much lower. The cost of voice-enabled instruction is about 1% of live instruction. In the United States, the cost of education is about $66 per day, per student for instruction. This does not include the costs of higher ed, exec ed, or workplace ed, which are much greater.

With voicebots, it is possible to get the per-learner cost of education below $0.60 per day. In fact, it may be a lot cheaper than that. This isn't like a class where you need to have a fixed amount of time per lesson and subject. It can be quicker or slower, depending on learner needs. In lieu of ninety minutes of reading, the knowledge transfer could prove to be much more rapid. Maybe twenty minutes is all that is required — or perhaps only five minutes.

By having customized e-learning, people are put on their own paths. There are also no commuting costs or facilities costs. It is just much more efficient in many ways in terms of both cost and time. Plus, without the constant interaction of the tech, you don't know what you don't know — and you aren't pushing yourself as much.

Voicebots can narrow the skills gap in areas of math, reading, science, or other topics. For a small amount of money, we could solve some fundamental issues in the education system. Starting with the launch of our first course in March 2020, Brainsprays had helped 60,000 preschoolers learn how to read on Alexa. It was funded by proceeds from a similar business use case.

In total, we now have about 210 courses of which about 70% are focused on reading, 25% are focused on math, and 5% are focused on science. The vast majority of our courses are focused on reading, because it's the easiest thing. It's fundamental.

We have two goals: teach kids how to read by the time they get to Kindergarten and teach kids how to do the multiplication tables by the time they get to Kindergarten. These may sound like unrealistic goals, but with voicebots technology, these goals are reasonable and highly achievable. ~

Lorenzo Carver is Co-Founder of Brainsprays.com, a smart speaker preschool that teaches 2- to 5-year olds to read and multiply like 7- to 10-year olds.

THE MOW-BOTS ARE COMING

Martin Buehler

- CTO of Wavemaker Labs and Product Manager of Graze -

Commercial mowing with large, autonomous mowers is on the verge of becoming a widespread reality. The landscaping industry is large at almost $115 billion per year in 2021 for the commercial landscaping industry. Plus, the landscaping industry, which employs over 1 million people, has seen fairly little automation at the commercial level to date. But the size of the industry isn't the only reason a half-dozen robotics companies are pushing to develop commercial automated mowers at scale. Financial and environmental fundamentals are key drivers pushing this technology ahead rapidly.

Landscaping is a sector with huge demand, poor employee retention, low skill requirements, and the job is not very exciting. Landscape companies have been having tremendous difficulty filling positions recently — especially against a backdrop of an aging workforce, a general labor shortage, and widespread rising wages.

Commercial mowing is also a dangerous business. And the mowing portion of landscaping is a loss-leader for many landscaping companies looking to secure larger, more complex and lucrative projects. In other words, this is a low-margin business. But with the use of commercial mow-bots, it is a sector that could be converted to a high-margin business.

It has become an issue of such magnitude because landscapers want more work. But they don't have people, so they can't take the jobs. This automation helps landscapers grow their businesses. They can also take on higher skilled workers to supervise and service the robots.

Commercial mowing is also a dirty business. Many of these vehicles are high emissions products. The emissions from gasoline mowers have become a major issue. In fact, the sale of gasoline-powered mowers is set to be banned in California in 2024. This makes the electric imperative for commercial mowers both time-bound and essential. Commercial landscaping companies as well as various communities are panicking. Time is short and requests are already coming from communities for our robotic mow-bots. By switching to automated, electric mow-bots, emissions, fumes, and noise will all be greatly reduced.

The mix of commercial and environmental dynamics lead us to believe that big things are coming for commercial autonomous mowing, including both autonomous and electric vehicles. In short, the mow-bots are coming.

Commercial mowing is a perfect job to automate — the economics, market dynamics, safety, and emissions align for this job to experience wide-spread automation. Mow-bots make commercial mowing better, higher paying, more productive, more profitable, safer, and greener. Robots that are electric can run day and night. And they do so quietly. Plus, the cost of electric mowing operations is a lot lower. Electricity costs are much lower than gasoline. Maintenance costs are also sharply lower.

Incorporating AI with mow-bots can have many other benefits. Mow-bots can better track and monitor timing and location of mowing, grass height, growing time, water needs, fertilizer needs, and many other vital things, like sprinkler malfunctions or even missing trash cans in parks. This is a good thing.

Not only will mow-bots bring benefits because they are being incentivized by regulation, health, safety, environmental goals, but they will also bring benefits, because data is at their operational core.

Why Are Mow-Bots Coming Now?

It's all about technological feasibility. It is now feasible to implement the use of commercial mow-bots at scale. It has been a difficult challenge. After all, there are three technological levels of challenges we have had to overcome.

First, hardware is hard. It's generally much harder to develop and create new hardware than software. Second, robotics is harder than most hardware, because robotics requires solving both hardware and software challenges, and their integration. Third, outdoor robotics is harder than indoor robotics. This is because hardware and software are forced to operate in unstructured and uncontrolled environments. Mow-bots need to deal with uneven and hilly terrain, slippery mud, rain, fog, wide temperature ranges, seasonal changes, and changing lighting conditions throughout the day. Weather, seasonal, and lighting issues present myriad challenges for vision sensors. Tree coverage may obstruct GPS and cause the robot's sense of location to fail. And everything needs to be weatherproof.

Graze Mow-Bot using electric and automation technologies.

While the mow-bot hurdles enunciated here have presented significant challenges, it is now feasible to overcome these issues, because the tech has improved. Sensors and sensor fusion algorithms are better. Plus, the software and autonomous technology stack that is available now makes things a lot easier. If you want to deploy this software, a lot more people know the necessary technology stacks. These factors make it easier to tackle this challenge now.

Given where the labor force is right now, there are a lot of financial incentives to implement mow-bots. But it is still challenging because it is very expensive to develop. You need high-pay, high-demand engineers. This is why you have to go for large markets with big opportunities and ROI. It's also why a half-dozen companies are working on this.

Wavemaker Labs is unique, which is why we are likely to be successful. Wavemaker Labs is a venture studio that founds startups in food and ag sectors, taking technology from "seed to fork." This is a pun on the notion of seed-stage funding as a dual meaning of the word seed. We found and fund our startups using equity crowdfunding that allows the general public to participate in our early-stage funding opportunities. We pretty much always have crowdfunding campaigns going at one time or another. For Graze, the company in Wavemaker Labs working on mow-bots, we have raised over $9 million so far. Equity crowdfunding is here to stay and there are particularly significant crowdfunding opportunities for consumer-interacting robots in the B2C area. Of course, it can work with B2B robot technology entities like Graze.

In just a couple years, electric mowers won't be optional. With an exceptionally tight labor market, autonomous mow-bots will be required to satisfy that massive demand. Government regulations are forcing the issue, the crowd is funding the push of development, and free market dynamics present strong ROI dynamics. In 2022, mow-bots are coming — because by 2024 they will be the only solution. ~

Martin Buehler is the CTO of Wavemaker Labs. He is currently serving as the product manager for the Graze portfolio company of Wavemaker Labs.

HUMAN LEADERSHIP IN THE AGE OF ROBOTICS, AUTOMATION, AND ARTIFICIAL INTELLIGENCE

Shannon McGurk

- Futurist with III Marine Expeditionary Force -

All good things have their price. The wonders made possible by automated management in the age of Robotics, Automation, and Artificial Intelligence (RAAI) are no different, but it is naive to think that we have not been influenced by the powerful engines of efficiency we have put in motion. These costs are likely to become more apparent in the years ahead, including in 2022.

As we continue to use RAAI to improve efficiencies in our economies, are our machines changing us in ways we cannot see?

How might our interaction with our machines and AI be affecting how we relate to each other? Likewise, what of the future? What hidden prices might we pay tomorrow for efficiencies we realize today?

In this article, I would like to suggest three things:

1. Interacting with technology in increasingly efficient markets shapes us,
2. Interacting with markets using no other criteria than efficiency objectifies us, and
3. Realizing a future we want in the face of RAAI-driven efficiency demands we do three things RAAI cannot do: *envision* a future we want, *lead* in an explicitly human way and *love* in an explicitly human way.

How Do We Know Markets Shape Us?

Common sense tells us our actions shape us. Purchasing a product or service provides signals to the market. We say what we value. We say how we decide. How fast and how often we buy signals how much we trust the process.

When anxious, we purchase and pay for services to reassure ourselves. We extend our credit when we feel either optimistic about the future, desperate about the present, or are exhausted by the complexity of the world. And when the market responds to *us*, it strengthens our actions, gratifies our faith, and confirms our behavior.

Drilling down deeply tells us *why* we decide and what it says about our motivations and values, but the powerful forces of advertising and the reinforcing mechanisms of the market both shape us and are shaped by us. Present Biasing, when we settle for a smaller reward now rather than a bigger reward later and Hyperbolic Discounting, when we show we prefer something sooner rather than later are two examples of many.

In their relentless drive toward efficiency, RAAI-driven markets present people as perfect objects. This presentation of people as objects tends to take two forms: people in images and people in idealized stories. While images objectify more than stories, both instances reduce human beings to objects. Objects are what we use, manipulate, and ultimately discard. We see this discarding dynamic casting an increasingly dark shadow over interpersonal relationships.

Envisioning Possible Futures: Two Ends of the Spectrum

What futures do increasingly efficient markets offer? At one end of the spectrum, maximized efficiency offers us further centralization of resources and a reduction of meaningful choices in an environment of machine-driven labor and market efficiency that objectifies human beings until we become absorbed by that efficiency.

This is the path we are on. On this path, we risk RAAI becoming our efficient master and we its commoditized servants who may ultimately become enslaved in the process.

At the other end of the spectrum, is a future of maximum decentralization of resources and a proliferation of meaningful choices in an environment of no less machine-driven labor and market efficiency that *subjectifies* human beings — keeps us the subjects in the relationship, not the objects — as we master efficiencies we realize. On this path we earn the privilege of receiving the benefits of RAAI, and remain its masters.

The fulcrum on which this spectrum of two futures rests is the difference between management and leadership. *Good management* — maximizing efficiency of known variables — can be realized by RAAI-enabled machines without us. *Good leadership* — identifying a better future and inspiring others to confront unknowns to achieve that future — demands our involvement.

Leadership in the age of RAAI is a values-driven process informed by three gifts only human beings can present: envisioning a future in which we respect the sanctity of the individual, asking better values-based questions, and serving in an explicitly human way.

Serving in an Explicitly Human Way

It is no coincidence that the United States has dominated the post-World War II global economy. While the U.S. has its detractors, the Rule of Law, free market capitalism, and the sanctity of the individual have worked together to help make the U.S. model the most successful in the history of the world.

Much of this success hinges on the sanctity of the individual, for the western model posits that there is a spark of the divine in each human being.

Inherent in the dignity of the individual are the freedoms many countries now take for granted. The Declaration of Independence, one of the founding documents of the United States, lists the fundamental rights of the individual as life, liberty and the pursuit of happiness.

Finally, while at its founding, the United States did not make sufficient allowance for gender equality, the equality of men and women is now foundational throughout the world. It is interesting that this fundamental equality has grown out of the political legacy of the United States' founding documents.

If sanctity of the individual, the individual's inherent right to freedom, and the equality of men and women are three values that contributed to the growth of the global economy, these are values worth promoting. A good way to promote these values is to ask better questions regarding possible threats to them.

While it may seem out of place to speak of love while examining RAAI, placing ourselves at the service of others is in fact a very logical way to function in a free market global economy enabled by RAAI. If we believe in the foregoing values, a question worth asking is, "What is the best way to take action in a market to promote the sanctity of the individual, to promote a free market where individuals can exchange goods and services as equals, whether they are men or women?"

Love in the Market: Serving in an Explicitly Human Way

Love is a discipline, not an emotion. This approach to love has fallen out of favor in the West, where we have come to believe love is an emotion. This approach is problematic, for emotions change. Instead we should ask ourselves, "What is the best way to demonstrate love as a decision and a discipline?" Love is *willing* the good of the other. I have seen no definition of love better suited to market economies.

A good example of this in a practical sense is a fiduciary relationship between an investment adviser and his client. Fiduciaries put clients' welfare equal or superior to their own. This dedication to service, is, in fact, love in the marketplace. Putting others before ourselves is counterintuitive and contrary to the efficiency of markets. This is precisely why it requires visionary ethical leadership.

Envisioning, Leading and Serving

Choosing a preferred future is a values-based decision. Where we want to go, what we want to do, and what we want to become mean choosing an ideal based on resources available and values we cherish. Leading is a decision to confront the unknown, in itself a values-based decision, but how we face the unknown also tells us a lot about ourselves.

Deciding to shape the future unknown based on confronting fear of the unknown is an endorsement of courage, itself a virtue. However, of the three actions, it is loving in an explicitly human way that reveals the most about ourselves.

Decentralized Finance: The Next Big Thing in 2022 and 2023

My prediction for the biggest change when it comes to AI and automation is the rise of Automated Market Making (AMM), smart contracts, and Distributed Ledger Technology (DLT). The world of Decentralized Finance (DEFI) started out a few years ago on the blockchains of Bitcoin and Ethereum, but this technology has exploded in recent years across multiple competing chains and is even being experimented with at the nation-state level, especially in China.

Here again, I see two distinct paths forward. The first is the path of China, and is a world where the state has absolute control and surveillance over every single monetary transaction on its own Central Bank Digital Currency (CBDC). This ultra-centralized model has the benefit of speed and efficiency, but is weak to tyrannical rulers, and can easily stifle innovation and freedom.

The second path is one of decentralization, building a world where Decentralized Autonomous Organizations (DAOs), AI, and smart contracts run most of the financial system, taking power away from the few ultra-wealthy and putting more power into the hands of the people.

Both of these models exist now, and both will see massive adoption over the next two years. It is up to us to decide whether we want a future of individual freedom and uncertainty or submit to a life of predictable oversight.

We need to choose wisely. ~

Shannon McGurk FLTA® is a China Futurist and Long-Term Analyst currently working in Okinawa with III Marine Expeditionary Force (III MEF). As a U.S. Army Ground Cavalry Officer and China Foreign Area Officer, he served on the Army and Joint Staff, with the Defense Intelligence Agency and at the U.S. Embassy in Beijing. As a contractor he has worked throughout the intelligence and operational communities, most recently in Afghanistan and with Air Force Futures and Concepts.

Disclaimer: The views and opinions expressed in this essay are those of the author. No U.S.M.C, Department of Defense, or Federal government endorsement should be inferred or implied.

INNOVATING TOGETHER FOR A SUSTAINABLE FUTURE

Anton Larsson

- Product Manager for Embedded Software & Data Science at Kollmorgen -

We are no longer settling for unsustainable. More than ever before, we are painfully aware that our ability — or inability — to act over the next few years will directly shape the lives of all future generations. And our best bet to step up to the challenge lies among the innovators — the ones who can adopt and apply new technology to truly transform the core of our society and create a world where we can thrive without the expense of our children.

A Market in Rapid Change

I have always been an optimist, as long as I can remember. Even as a kid, I had a strong belief in people's ability to overcome hard challenges and, when it mattered most, do the right thing.

I think that is what attracted me about Kollmorgen — the opportunity to work for a company that so clearly reflected such values. It is a company that wants to make the world a better place by unlocking the full potential of innovators; and furthermore, being able to do so in a market with so much potential.

Because there is no doubt that the future of mobile robotics is looking bright, clearly reflected by rapid growth rates. The COVID-19 pandemic spurred demand even further, and the rules of the game are changing faster than ever before. That's also why identifying "the one big thing of 2022" has been a bit of a challenge.

However, emerging from of a worldwide pandemic tends to change your perspective a bit. For one, it humbles you and makes you realize how vulnerable you are. But for me personally, it has also strengthened the optimist within me. By being able to observe how we as a global society have come together, stayed patient, and restructured, in order to beat COVID-19, I believe that we are now ready to address the real elephant in the room, the greatest challenge of them all: environmental sustainability and saving our planet.

The Shift to Sustainable

In late 2021, global leaders gathered for two weeks at the United Nations Climate Change Conference in Glasgow to align on how to mitigate climate change. Although some would call the outcome disappointing and the established goals set up not aggressive enough, I believe that there is light at the end of the tunnel.

Because the trend is clear: there is a more powerful force in motion, one that has the potential to really make a difference. I am referring to *The Market.*

Data from Forrester Research indicates that people are more aware of the condition of the environment than ever before, more willing to take action on environmental sustainability and actively seeking brands and products that are truly committed to sustainability and consider it through their entire supply chains. Investment companies are now only willing to invest in sustainable enterprises, not only because it is the right thing to do, but because it's the safest long-term investment to make from a monetary perspective.

Corporations all over the world are quickly making sustainability core of their strategies, because it is *no longer just a matter of ethics and responsibility, it's a matter of money.*

Empowering the Innovators

Industries generate roughly a quarter of global emissions annually and it is critical to drastically reduce this number in order to mitigate the worst effects of climate change. Mobile robotics has the ability to lead the development in more sustainable industries and help companies all over the world meet their sustainability goals. Just consider data from Statistica that clearly reports how mobile robotics will not only significantly increase energy-efficiency and productivity of industries that adopt it, but also minimize errors, and thereby waste. We see first-hand examples of this every day at Kollmorgen.

But the real question is not whether mobile robotics has the possibility to create more sustainable industries, but rather how we can scale it fast enough? How can we accelerate adoption that would normally take decades?

At Kollmorgen, the answer is clear: By enabling and empowering the innovators! We need to bring together the sharpest minds of the industry and give them the right tools to solve the problem. Cutting-edge technology will no doubt play key role.

We see clearly how AI and machine learning will have the ability to drastically reduce the complexity of deploying a mobile robot. Many of the tasks that until recently required weeks or months of work from experienced and skilled engineers will soon be performable in minutes by algorithms. And a lot comes down to data.

By providing innovators with data and insights from over 50 years in the business and more than 40,000 deployed Kollmorgen -based mobile robots, we can create smarter solutions that put mass-scaling of robotics within reach, bringing greater sustainability to industries all over the world.

Just as important will be openness and creating the opportunity for collaboration. Achieving truly sustainable industries means deploying the most suitable robot for each application and ensuring that they operate seamlessly together in the same system. Not an easy task by any means. But a collaborative approach is key to success!

At Kollmorgen, we believe that means supplying an automation platform that is modular and easy to build applications and integrations on. We invite OEMs all over the world to innovate together with us, and we provide the tools, experience, and insights within fleet management, navigation, and mobile robot deployment — to let our partners focus on their core expertise.

A Call for Action

Driven by an irreversible shift in the market, fueled by the optimism of knowing what we are capable of when we put our hardest effort into it, I believe that — at one minute before midnight — *sustainability* will finally get the full attention it needs and deserves to become the "one big thing" for technology, industry, and the market in 2022.

Kollmorgen AGV control systems.

For the innovators who can step up to the challenge and find a way of scaling mobile robots across the world, it will not only be an attractive business. They can rest assured that they are leaders on the path to a more sustainable society. Because in the end, our time on earth is limited, and there is a lot of truth to that old saying, "We do not inherit the Earth from our ancestors, we borrow it from our children." ~

Anton Larsson is the Product Manager for Embedded Software & Data Science at Kollmorgen.

REFINING TECHNOLOGY IN PROJECT MANAGEMENT

Nawfal Patel

- Program Manager at Google -

The last two years have caused all businesses to reevaluate and adapt how they perform their operations. The forced shift to remote work from COVID-19 resulted in a rapid embrace of collaboration and project management technologies. Project managers have had to embrace technological solutions to work with distributed teams and unexpected challenges. There has been greater emphasis on soft-skills for project managers as they work to bring the best results out of their team while working remotely. However, the biggest change in project management in the next year will be the integration of increasingly complex and custom technologies that suit individual and business needs.

Project managers are needed in every industry, but depending on the type of projects they manage the technology they are looking for might require different things.

Technical project managers might have more need for task management and flexibility whereas a manager in manufacturing could be looking for more progress tracking and performance mapping features.

Overall, most businesses manage projects with a hybrid agile methodology that allows them to be flexible when necessary. Affordable agile project management tools such as Slack and Asana have seen increased adoption by small businesses as leaders shifted to the new normal after the COVID-19 pandemic. Remote work is here to stay with 74% of workers saying having the option to work remotely would make them less likely to leave a company.[1] Using team collaboration systems for workflow management among remote teams will only increase in the future.

However, as more businesses began to integrate third-party project management technologies into their existing operations, a clear gap emerged for some where the technologies didn't exactly suit their needs. This led managers looking to implement new solutions into their processes to seek out customizable project management technologies to work best for them. More complex project management tools like Atlassian JIRA and Trello have seen significant growth because they provide far greater customization and scaling options. Some larger corporations have even developed their own project management solutions for internal use. Google has custom built internal project management tools called Buganizer and Taskflow which employees use to manage tasks and track progress.

In 2020, a report revealed that 54% of organizations lacked access to real-time KPI's for their projects and over a third of them spend more than one day collecting data.[2] Increased data collection allows for company-specific key performance indicators that can enhance collaboration, especially among distributed teams.

Poor management of resources is one of the biggest issues across the project management industry. Managing people with technology can get complicated and understanding exactly how a team is performing throughout a project will help businesses to meet their goals. This is why companies are turning to more advanced applications that provide increased resource visibility and management.

One of the biggest benefits of integrating new project management technologies with automation for a business is the ability to track project progress. The automation of progress tracking can provide organizations with much more data and insight into the current status of a project as well as the potential outcomes, both good and bad. Artificial intelligence has also begun to play a role where it can, such as, analyzing potential risks, managing the allocation of resources, and leveraging historical data to plan for future projects.

As these technologies are developed further, artificial intelligence will become a much more important part of the project management process.

Instead of a human planning a project and manually monitoring progress, an AI system could be programmed to use past results to create a project plan, monitor progress in real-time, and provide schedule or budget updates to management. This means that project managers can be extremely adaptable, react quickly to any issues that pop-up, and keep stakeholders well informed throughout a project's life-cycle. It will also save a significant amount of time that can be used in other ways.

While project management as an industry has been keen to embrace technology, a more selective approach might work better in the year ahead and beyond. As we look forward, businesses will need to empower their project managers to seek out technological solutions that can be used most effectively for them. Integrating automation can be tricky and will need to be managed carefully. Different organizational structures and process flows make using a one-size-fits-all solution difficult. Project managers will need to balance their desire for efficiency while recognizing the difficulties in overseeing a remote team to achieve the best results and keep their team members happy. ~

Nawfal Patel PMP® FLTA® is a program manager in Google's Privacy and Data Protection Office. He is the former operations manager of The Futurist Institute.

ENDNOTES

1. https://www.victoryworkspace.com/101-statistics-about-remote-working-in-2021/
2. https://wellingtone.co.uk/wp-content/uploads/2020/06/The-State-of-Project-Management-Report-2020-Wellingtone.pdf

SEEING SUPPLY CHAIN AS A SYSTEM AND FINDING INSIGHTS IN DATA

Jim Rice

- Deputy Director at
MIT Center for Transportation and Logistics -

No one wants to talk about port capacity and the supply chain when it's not an issue. But in some respects, supply chain operations are like oxygen — it's something you really miss when you don't have it. When it is an issue, this bubbles right to the top of the priorities list. That's exactly what happened in 2021, and we enter 2022 with these systemic challenges as well as a growing potential for the supply chain to overcorrect in response to pervasive shortages.

Seeing Supply Chain as a System in 2022

Is 2022 the year when people wake up to the notion of supply chain as a system? That's called a Christmas Wish. It's really exciting that so much attention is being paid to supply chain — a thing we know and love. But it is also something that so many people still don't understand just yet.

What problem solvers — especially business leaders, executives, and policymakers — need to understand is how supply chain operates at the system level if they genuinely wish to create solutions that are effective. For example, it's important to understand how things like inventories cover over sins in the system. There may be port constraints, factory outages, and surges in demand. But as long as inventories are high, those issues can be glossed over, invisible to the public.

Recently, with the transportation and supply chain systems at capacity and stressed, inventories have diminished to record low levels. This has now become a global and high-impact problem, with the limitations of the supply chain system exposed for all to see. After all, once the inventories are gone, all you have left is the system. To put it another way, if you want to know where the rocks in the ocean are, empty the ocean. That's what's happened in the supply chain. We've removed the water and all we see are rocks.

On the upside, volumes are way up. Supply chains and transportation systems are producing and moving more goods than ever before. And despite the focus on clogged ports with lines of ships waiting to offload their cargo, record levels of cargo were offloaded in October 2021 in the United States. The problem is that even offloading a record level of cargo has not been enough to process all the existing demand. The supply chain is only so big and has only so much capacity.

With depleted inventories, constrained operations, and material and labor shortages, supply chains are at their capacity limit. And when operating at capacity, systems are hypersensitive to even small disruptions. The result — widespread delays, and people don't get what they want when they want it. One small example might help to illustrate this in action.

Right now, warehouses are working at high capacity and throughput volumes are very high. This is a problem in warehouses because if you fill up a warehouse, you have no space to actually receive goods, store them, and then pull them out to prepare for a shipment. This is called honeycombing and when that happens in a warehouse, throughput volume drops significantly. We've seen a similar dynamic in ports, which have very little unused space to offload cargo from vessels and to stage cargo for loading.

Consumers and companies respond in several ways — stocking up and overordering — which can make matters worse. We at the Center for Transportation and Logistics run an exercise for our executive and graduate students which highlights these phenomena known as the "bullwhip effect." Small changes upstream lead to big problems downstream and the system structure leads to increasing and alternating shortages, excess inventory, and delays. We're in the middle of those dynamics right now, and we can predict that our current situation of massive shortages will likely be followed in the not-too-distant future by oversupply, bloated inventories, and some clearance sales. It is very difficult to match supply and demand when there are no inventory buffers to allow for variation.

That being said, as we look ahead to 2022, the question we need to ask is, what can we do better this time? My answer is that we can use data to understand how the supply chain performs, and we can use that data to better inform our strategic decision making to redesign supply chains to avoid the bullwhips and excess supplies that often follow periods of shortages.

Letting Data Guide Strategy in 2022

Data availability and use is increasing. Looking ahead to both 2022 and beyond, more information will be created, and hopefully that information will be used to inform decision making. I would expect to see the use of data grow for solving point problems, but also for supply chain design. Data interpretation and the software that facilitates it have improved, become more available, and are offering more value to decision markets. More data can help inform machine learning and artificial intelligence programs, and properly deployed, could provide an improvement — if not step change — in decision support. That will be relevant for supply chain training and education in the year ahead and beyond.

We teach our graduate students to use models to help inform decisions, and we are more actively using visualizations to help broaden the use of models. This is especially helpful to illustrate key tradeoffs without requiring that one become an expert in a modeling tool. Looking ahead, we believe there will be more and greater use of visualizations for executives, especially recognizing the importance of understanding the supply chain as a system.

Data visualizations can help executives appreciate the tradeoffs in supply chain design and can facilitate productive discussions about those tradeoffs with the key decision makers. And supply chain design will be critical for organizations in the future, in part because the pandemic has surfaced existing supply chain system limitations.

Applying the Data to Not Overdo It

Given these revealed system limitations, data can likely add value when considering the supply chain as a system. Demand data can obviously be useful for adding transparency and helping decision-makers predict when the system may see a reduction in demand, but also help identify constraints as in the current situation, when the supply chain has been overextended. Some companies, as noted above, are operating in overdrive and adding capacity. Adding capacity, however, doesn't always make economic sense. For some companies, such as for a company in a capital-intense industry with low margins, high utilization, and low growth (e.g., toilet paper), adding capacity would only make sense if there were a sustained increase in base demand or a significant loss of base capacity in the marketplace.

Even in cases where adding capacity does make economic sense, the cycle and timeline to add capacity can be years and cost billions of dollars, which is the case for semiconductors. This means that decisions to add capacity won't provide much relief in the near term. Semiconductor manufacturers have made decisions to add capacity based on some predicted increase in base demand, but most of that added capacity will not provide relief for the current dilemma.

The Broken Supply Chain?

In the media, people say the supply chain is broken. Maybe it is, but any system stressed beyond its normal capacities will suffer similar problems. In any system, there will be bottlenecks, and often when you remove one, it will reveal another bottleneck elsewhere in the system. It's like that carnival game of whack-a-mole where you have to hit the mole that pops up on the board, and once you hit one, another immediately pops up elsewhere. The moles just keep popping up.

In supply chain terms, we only notice the bottlenecks (the moles) when demand surges and the bottlenecks are tested and exposed. When we address one bottleneck, another one surfaces (another mole) and so the problem moves around. We are now seeing port congestion, but also constraints in transportation and warehousing as well.

Currently we are hearing a lot about storage and warehousing issues — these illustrate the problem of exceeding capacity and moving bottleneck problem. When warehouses utilize all of their space, it becomes difficult to move around and do work with goods stacked everywhere. This honeycombing results in added delays, big inefficiencies and higher costs. This occurs in ports as well, where stacks of containers make it very difficult to load and unload vessels. In order to load a vessel or an outbound truck, terminal operators need to find the desired containers and often disassemble stacks of containers to get to the things they need. To operate these facilities efficiently, one can't fully utilize the space for storage alone — you need space to actually move the materials.

When there was inventory and additional response capacity in the systems, we never really exposed the bottlenecks and therefore didn't experience these kinds of delays. Clearly, many in the public domain have taken an interest in the supply chain, and some are wondering how it can be fixed. This leads to asking how much capacity our supply chain can handle, and these are key supply chain design questions that are made by each company.

At the national level, some may wonder whether we need an economy and transportation system that can handle the current level of capacity as its normal steady state. Maybe or maybe not, but this is not a simple situation with private companies making production and transportation capacity decisions and governments making infrastructure decisions. Even if it were simple (it's not), it would be very costly to increase the overall capacity of the global transportation system and supply chain.

There isn't one solution to the current set of problems — there are many interconnected systems, each of which may require a distinct solution. As noted above, the cycle time to add capacities varies greatly. For example, regarding the transportation delays — it would take years and years to add a domestic port, but it would only take months to add actual trucks. Of course, you need truck drivers too — so critical today. Adding skilled truck drivers requires investments in compensation, job design, and training — enough to make the profession more appealing than it is right now. But other resources are required too — more containers, more chassis, potentially more port capacity.

Given that adding a port would take many years, leveraging existing capacity of ports by making them more resilient would be important. As suggested above, the current system of domestic U.S. ports obviously has capacity constraints and current congestions are revealing this. At the moment, there are many constraints, including those imposed by the Jones Act and the physical constraints of existing ports that do not have the ability to handle the largest container ships today. All of this stresses the system to the point that if you lose one major port, there are significant risks to the entire system. There just isn't enough redundancy in the domestic U.S. port system, so when you lose one major port, delays will ensue, and critical supplies will need to find alternate transportation methods to get to their customers.

Using Data to Make the Right Investments

Hopefully, we can make data-informed decisions about capacity. This includes decisions about capacity of existing assets as well as the decisions to add productive assets. Not just for predicting the inflection points of growth and demand, but also for thinking about the capacity at the business level to continue to be profitable and productive.

Investments in the system ultimately will need to be guided by broader strategic decisions about several core capabilities. Two key capabilities are resilience and sustainability. Redesigning a supply chain involves making choices about throughput capacities, but also about how your supply chain will be able to respond to disruptions.

There are seven core capacities to create resilient supply chains, and these ought to be considered and designed into the supply chain. At the same time, designing the supply chain to be sustainable is important for the future viability of operations and our environment. These are emerging capabilities that will be increasingly important in the future.

There is a tenuous connection between investments in these and bottom-line profitability which makes these challenging for organizations to embrace and support. But the good news is that there is work being done to make the business case clearer for supporting these investments, and hopefully in 2022 we will see some progress in this area.

Automation in the Supply Chain

With the labor market stressed, one area that is growing and has the potential for more growth is automation. As we think about widespread use of robots, cobots, and automated vehicles across the supply chain, these are in early stages of application and deployment. But there is an ongoing and evolving set of dynamics that will continue to play out in incremental ways which appear to encourage the increased development and adoption of automation for an increasing array of applications. Given that, one needs to be thoughtful about the practical use of automation in the workplace alongside human beings — both the executives and front-line supply chain staff — who will need to be trained in developing safe and productive applications.

The goal of incorporating and integrating automation into supply chain operations is to get the best of both worlds, with humans processing situations qualitatively and robots (as well as AI and other automated solutions) taking advantage of lots of data for quantitative real-time use.

The Year of the Breakdown?

Is 2022 the year of supply chain breakdown? I don't think so. We've experienced delays through late 2021 and this may continue, but I think some relief will come because manufacturing and supply chain demand is not likely to maintain the recent pace.

While we haven't had a COVID-like experience of demand and supply spikes until now, we have had an overheated economy before. This is why business cycle data can also be very useful for helping companies see the big picture across the system to hopefully be better able to predict when we may see a slowing of demand. While the origins of current supply chain dynamics may be unique, permutations of current challenges are not unique.

As we look to 2022 and beyond, executives, strategists, and policymakers will be addressing many different challenges within the supply chain. The solutions are likely to be resolved by addressing multiple issues of capacity and response throughout the supply chain. Each node in the supply chain will need to address the questions of supply chain design (how much capacity, how much resilience, how sustainable, etc.). It will also call for the government to determine the desired capacity for public infrastructure like the highways and port systems.

These are tough questions to address, and they require a lot of deep analysis and supply chain system design. As companies rush to solve supply chain issues and meet strong demand, we can hope that they are also looking to these big questions to deal with future disruptions that will most certainly be coming. ~

Jim Rice is the Deputy Director of the MIT Center of Transportation and Logistics. Launched in 1973, the MIT Center for Transportation and Logistics is a dynamic solutions-oriented environment where students, faculty, and industry leaders pool their knowledge and experience to advance supply chain education and research.

FI THE FUTURIST INSTITUTE

The Futurist Institute was founded in 2016 to help analysts, executives, and professionals incorporate technology and trend risks and opportunities into strategic planning. The Futurist Institute confers the Futurist and Long-Term Analyst™ FLTA® designation and helps analysts become Certified Futurists™. Our courses have been approved for continuing education hours by the Certified Financial Planner Board of Standards (CFP Board).

Current Courses

The Future After COVID
The Future of Work
The Future of Data
The Future of Energy
The Future of Finance
The Future of Healthcare
The Future of Leadership
The Future of Transportation
Futurist Fundamentals
Quantum Computing
Write the Future

Visit The Futurist Institute:

www.futuristinstitute.org

ABOUT THE EDITOR

Jason Schenker is the President of Prestige Economics and Chairman of The Futurist Institute. He has been ranked one of the most accurate financial forecasters and futurists in the world. Bloomberg News has ranked Mr. Schenker a top forecaster in 46 categories, including #1 in the world for his accuracy in 26 categories, including for his forecasts of the Euro, the British Pound, the Russian Ruble, the Chinese RMB, crude oil prices, natural gas prices, gold prices, industrial metals prices, agricultural commodity prices, and U.S. jobs.

Mr. Schenker's work has been featured in *The Wall Street Journal, The New York Times,* and the *Frankfurter Allgemeine Zeitung*. He has appeared on CNBC, CNN, ABC, NBC, MSNBC, Fox, Fox Business, BNN, Bloomberg Germany, and the BBC. Mr. Schenker has also been a guest host of Bloomberg Television, and he is a contributor for *Bloomberg Opinion*. Mr. Schenker was ranked one of the top 100 most influential financial advisors in the world by Investopedia in 2018. He is a Governance Fellow of the National Association of Corporate Directors and a member of the elite Bretton Woods Committee.

PUBLISHER

Prestige Professional Publishing was founded in 2011 to produce insightful and timely professional reference books. We are registered with the Library of Congress.

Published Titles

After the COVID Vaccine
Be the Shredder, Not the Shred
Commodity Prices 101
Electing Recession
Financial Risk Management Fundamentals
Futureproof Supply Chain
A Gentle Introduction to Audit and Due Diligence
Jobs for Robots
Jobs for Robots: COVID Edition
Midterm Economics
Quantum: Computing Nouveau
Reading the Economic Tea Leaves
Recession-Proof Career Strategies After COVID
Robot-Proof Yourself
Spikes: Growth Hacking Leadership
Strategic Cost-Cutting
Strategic Cost-Cutting After COVID
The Dumpster Fire Election
The Economics of Revolt and Revolution
The Fog of Data
The Future After COVID
The Future of Agriculture

PUBLISHER

Published Titles

The Future of Energy
The Future of Finance is Now
The Future of Food and Agriculture After COVID
The Future of Travel
The Promise of Blockchain
Write the Future
The Robot and Automation Almanac - 2018
The Robot and Automation Almanac - 2019
The Robot and Automation Almanac - 2020
The Robot and Automation Almanac - 2021
The Robot and Automation Almanac - 2022

Forthcoming Titles

Content Monster
Disruption Warfare
The Author Institute Method
The Future of Healthcare
The Future of Marketing

JOBS FOR ROBOTS

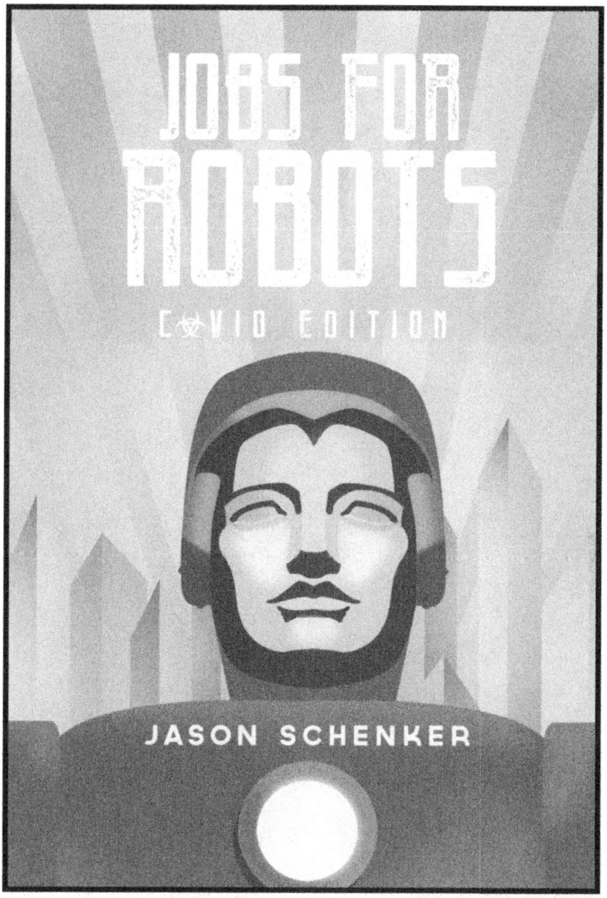

Jobs for Robots: COVID Edition provides an in-depth look at the future of automation and robots, with a focus on opportunities and risks for the future following the COVID-19 pandemic. The first edition of *Jobs for Robots* was published in February 2017. It has been a #1 Best Seller on Amazon.

THE FUTURE OF TRAVEL

In *The Future of Travel: Trends and Technologies Shaping the Decade Ahead* from The Futurist Institute, leading experts, futurists, and executives in travel, leisure, and hospitality share essays focused on the most important future trends impacting these industries, including the associated opportunities, priorities, perception shifts, and challenges shaping the decade ahead.

DISCLAIMER

FROM THE FUTURIST INSTITUTE

The following disclaimer applies to any content in this book:

This book is commentary intended for general information use only and is not investment advice. The Futurist Institute does not make recommendations on any specific or general investments, investment types, asset classes, non-regulated markets, specific equities, bonds, or other investment vehicles. The Futurist Institute does not guarantee the completeness or accuracy of analyses and statements in this book, nor does The Futurist Institute assume any liability for any losses that may result from the reliance by any person or entity on this information. Opinions, forecasts, and information are subject to change without notice. This book does not represent a solicitation or offer of financial or advisory services or products, and are market commentary intended and written for general information use only. This book does not constitute investment advice.

DISCLAIMER

FROM THE PUBLISHER

The following disclaimer applies to any content in this book:

This book is commentary intended for general information use only and is not investment advice. Prestige Professional Publishing LLC does not make recommendations on any specific or general investments, investment types, asset classes, unregulated markets, specific equities, bonds, or other investment vehicles. Prestige Professional Publishing LLC does not guarantee the completeness or accuracy of analyses and statements in this book, nor does Prestige Professional Publishing LLC assume any liability for any losses that may result from the reliance by any person or entity on this information. Opinions, forecasts, and information are subject to change without notice. This book does not represent a solicitation or offer of financial or advisory services or products; this book is only market commentary intended and written for general information use only. This book does not constitute investment advice.

Prestige Professional Publishing, LLC
4412 City Park Road #4
Austin, Texas 78730
www.prestigeprofessionalpublishing.com

ISBN: 978-1-946197-76-4 *Paperback*
978-1-946197-77-1 *Ebook*

THE FUTURIST INSTITUTE

Made in the USA
Monee, IL
04 November 2026

67387547R00105